Successful Implementation of Concurrent Engineering Products and Processes

Successful Implementation of Concurrent Engineering Products and Processes

Edited by
Sammy G. Shina
University of Massachusetts, Lowell

JOHN WILEY & SONS, INC.
New York   Chichester   Weinheim   Brisbane   Singapore   Toronto

This book is printed on acid-free paper. ☉

Copyright © 1994 by John Wiley & Sons, Inc. All rights reserved.

Published simultaneously in Canada.

No part of this publication may be reproduced, stored in a retrieval system or transmitted in any form or by any means, electronic, mechanical, photocopying, recording, scanning or otherwise, except as permitted under Sections 107 or 108 of the 1976 United States Copyright Act, without either the prior written permission of the Publisher, or authorization through payment of the appropriate per-copy fee to the Copyright Clearance Center, 222 Rosewood Drive, Danvers, MA 01923, (978) 750-8400, fax (978) 750-4744. Requests to the Publisher for permission should be addressed to the Permissions Department, John Wiley & Sons, Inc., 605 Third Avenue, New York, NY 10158-0012, (212) 850-6011, fax (212) 850-6008, E-Mail: PERMREQ@WILEY.COM.

This publication is designed to provide accurate and authoritative information in regard to the subject matter covered. It is sold with the understanding that the publisher is not engaged in rendering professional services. If professional advice or other expert assistance is required, the services of a competent professional person should be sought.

Library of Congress Cataloging-in-Publication Data:

Successful implementation of concurrent engineering products and
  processes / [edited by] Sammy G. Shina.
    p. cm.
  Includes bibliographical references and index.
  ISBN 0-471-28510-2
    1. Electronic industries—Case studies.   2. Concurrent
  engineering—Case studies.   I. Shina, Sammy G.
TX7836.S83   1993
621.38'068'5—dc20   93-8487

Printed in the United States of America

10   9   8   7   6   5   4   3   2

Contents

Preface   xi

Acknowledgments   xiii

Chapter 1   Concurrent Engineering in New Products and Processes   1

NEW PRODUCT TYPES   1
THE MATRIX ORGANIZATIONAL STRUCTURE   3
THE PRODUCT CREATION PROCESS   4
PROJECT MANAGEMENT MODELS   5
AXIOMS FOR CREATING CONCURRENT ENGINEERING WITHIN THE ORGANIZATION   6
EMPHASIZE THE TEAM FOCUS APPROACH TO PROJECT MANAGEMENT   10

Chapter 2   QFD Comes to Raychem Corporation: The Story of a Pilot Project   16

CHAPTER OVERVIEW   16
COMPANY PRODUCTS AND STRUCTURE   16
SELECTING THE PILOT PROJECT   18
DEFINING CUSTOMER REQUIREMENTS   22

DEFINING PRODUCT TECHNICAL
   REQUIREMENTS   26
ANALYZING THE COMPETITION   31
REFINING THE DESIGN   32
BEYOND THE HOUSE OF QUALITY   34
THE IMPACT OF USING QFD   36
INTEGRATING QFD INTO PRODUCT
   DEVELOPMENT PRACTICES   40
CHAPTER SUMMARY   42

Chapter 3   **Concurrent Engineering Delivers on its Promises: Hewlett Packard's 34401A Multimeter   44**

INTRODUCTION   44
MARKET BACKGROUND   45
HP34401A TEAM FORMATION   45
LEARNING TO WORK TOGETHER   46
THE TOOLS   46
A TOUR OF THE FINAL ASSEMBLY AREA   56
DEALING WITH NEW TECHNOLOGY
   AND ADDITIONAL CONCURRENT
   ENGINEERING   61
RESULTS DELIVERED BY CONCURRENT
   ENGINEERING   62
SUMMARY   63
ACKNOWLEDGMENTS   64

Chapter 4   **A Team Approach to Concurrent Engineering: A Case Study   65**

INTRODUCTION   65
TEAM APPROACH TO PRODUCT
   DEVELOPMENT   67
PETRI NETS AS A TOOL
   IN PRODUCT DEVELOPMENT   68
SIMPLIFIED TIMED PETRI NET   69
PRODUCT DEVELOPMENT—
   HACH COMPANY CASE STUDY   70

INTERFACILITY COMMUNICATIONS 80
RULES APPLIED TO THE PETRI NET MODEL 83
WHAT THE STUDY SHOWED 84

Chapter 5 Concurrent Engineering at Chipcom: Implementation at a Small Corporation 88

INTRODUCTION 88
THE CORPORATION 89
THE CHIPCOM SMARTHUB 89
CHIPCOM CULTURE 90
THE CURRENT NEW PRODUCT DEVELOPMENT
   PROCESS 90
WHY CHANGE? 93
CONCURRENT ENGINEERING'S EARLY
   STAGES 94
THE PLAN 97
TOOLS 105
TRAINING 109
MEASUREMENTS 110
CONCLUSION 111
EPILOGUE 111
DISCLAIMER 112

Chapter 6 Concurrent Engineering: Sun Microsystems 113

PURPOSE 113
INTRODUCTION 113
CONCURRENT ENGINEERING APPLICATION
   IN THE NEW PRODUCT DEVELOPMENT 114
SUN MICROSYSTEMS NEW PRODUCT
   INTRODUCTION MODEL 114
SUN MICROSYSTEMS BUSINESS TEAM AND
   PRODUCT TEAM MODEL 115
ROLE OF THE NPI TEAM MEMBERS IN
   THE CONCURRENT ENGINEERING
   CAPACITY 117

PRODUCT PIPELINE 120
PRODUCT DEVELOPMENT CYCLE 120
CONCURRENT ENGINEERING EXAMPLE 1:
 COMPETITIVE ANALYSIS 120
CONCURRENT ENGINEERING EXAMPLE 2: DFM
 IMPLEMENTATION INTO NEW PRODUCT 123
CONCURRENT ENGINEERING EXAMPLE 3:
 PREVENTATIVE MEASURES TO ENSURE
 TIMELY PRODUCT AVAILABILITY 123

Chapter 7 From Specification to Beta Site in Three Months: Concurrent Engineering at Mercury 126

THE MERCURY ADVANTAGE 127
THE PRODUCT CREATION PROCESS 128
MERCURY'S MANUFACTURING
 PHILOSOPHY 128
SUPPLIER PARTNERSHIPS 134

Chapter 8 Design for Manufacturability at Northern Telecom 137

THE GATE PROCESS 138
THE NEW PRODUCT MODULE 139
THE MANUFACTURABILITY ASSESSMENT 140
THE EVOLUTION OF THE
 CORPORATE STANDARDS 145
RESULTS OF THE DESIGN FOR
 MANUFACTURABILITY EFFORT
 AT NORTHERN TELECOM 146
THE FUTURE CHALLENGE 147

Chapter 9 Concurrent Engineering of a Polaroid Camera and Flexible Automation 150

THE NEED FOR CONCURRENT
 ENGINEERING 150

THE APPROACH USING FLEXIBLE
    AUTOMATION   151
PRELIMINARY ENGINEERING FOR FLEXIBLE
    AUTOMATION   153
SUBASSEMBLY AUTOMATION FOR THE NEW
    CAMERA LINE   154
THE NEXT STEP—FINAL CAMERA
    AUTOMATION   159
SUMMARY   159

Chapter 10   JITQC: An Employee-Centered Methodology Designed to Impact Product Quality in an SMT Manufacturing Line   163

INTRODUCTION   163
METHODOLOGIES   165
DATA ANALYSIS   170
ANALYSIS   179
SUMMARY   183
RECOMMENDATIONS   184

Appendix 10-A Final Assembly Survey   186

Chapter 11   Improving Manufacturers' Distributor Performance Using QFD   189

PROJECT BACKGROUND   189
IMPLEMENTATION   190
ANALYSIS   192
PRIORITIZATION AND
    INSTITUTIONALIZATION   196
MONITORING/FEEDBACK   201
CONCLUSION   203

x   Contents

Chapter 12 Measurement System Analysis   206

MEASUREMENT SYSTEM ANALYSIS   209
ACCURACY AND DISCRIMINATION   210
REPEATABILITY   212
REPRODUCIBILITY   214
GR&R   217
LINEARITY AND STABILITY   220
CASE STUDIES   221

Chapter 13 The Use of Information System to Enhance Concurrent Engineering at MA/COM Omnispectra   226

HISTORICAL PERSPECTIVE   228
DOCUMENTATION REQUIREMENTS AND CONTROL   229
SYSTEM CONNECTIVITY   232
ENGINEERING CHANGE ORDERS (ECNs)   233
RESULTS OF THE INFORMATION SYSTEM INVESTMENTS   233

Chapter 14 Lessons Learned in Concurrent Product and Process Development   238

CASE BACKGROUND   239
PRODUCT VERSUS PROCESS   240
TECHNOLOGY VERSUS OPERATIONS   246
DEPARTMENTAL VERSUS SYSTEM VIEW   254
CONCLUSION   259

Index   261

Preface

Concurrent engineering is becoming more important as companies compete in a worldwide market. Successful implementation in different companies does not follow identical scripts, but the tools and methodologies of concurrent engineering are fused with the company's culture to create a unique and successful blend in each instance.

This book is a collection of case studies detailing successful implementation of the tools and methodologies of concurrent engineering at different companies. They vary in size, scope, product type, and strategy. Yet they are similar in their approach of successfully implementing concurrent engineering through an interdisciplinary team environment and in using the tools and methods effectively by altering them to meet their particular needs. They share a common vision of focusing on the new product creation and development process and have clear management support for the team effort and its mandate.

In many instances, the case study represents the company's first experience with concurrent engineering. Each is told by a principal member of the team, and it presents a brief chronology of the history and events that led to the successful implementation of concurrent engineering effort and documents its results. All the chapter authors are representing their teams and companies, and recording their success stories from their own experiences as they participated in the concurrent engineering effort. Both product and process development projects are outlined, and the case studies represent actual products and processes that met their original goals and objectives.

I hope that this book will be of value to the neophyte as well as to the experienced practitioner of concurrent engineering. In particular, I hope that the small to medium-size companies that do not have a large support staff and resources try out some of the concurrent engineering ideas and techniques and meld them into the company culture. The experiences documented here should be helpful in encouraging many companies to venture out and develop new world class products through concurrent engineering that can make them grow and prosper in the future.

Sammy G. Shina

Acknowledgments

The editor wishes to acknowledge the contributing chapter authors for their hard work and salute them for their contribution to the advancement of concurrent engineering. In many cases, the chapters represented their first effort at writing and publishing a major article. Engineers are well known for their lack of fondness for the written word, and it took many attempts at coaxing the authors in order to have them tell us their stories.

The editor also wishes to thank the many companies represented in this book, for allowing their employees to share with us their culture and their internal project efforts. Their contribution to advancing the state of the art in concurrent engineering is well appreciated, and the hope is that many other companies can benefit from and share in their success.

The editor is indebted to several organizations for supporting and encouraging him during the time it took to contact the authors, collect the chapters, and edit the book: The University of Massachusetts Lowell, for its continuing support for product design and manufacturing, especially Chancellor Bill Hogan's office; the Dean of the James B. Francis College of Engineering, Aldo Crugnola; and the chairman of the Department of Mechanical Engineering, Bill Kyros.

In addition, the editor wishes to thank Reed International PLC through their National Electronic Packaging and Production Conferences (NEP CON), for the opportunity to meet many of the chapter authors through their nationwide conferences and technical sessions. Special mention to the two principals in the Chicago division: Mike Critser, director of Conferences; and Sergio Alves, Conference Manager. Also, thanks to Philips Kommunikations Industrie AG, for providing a unique international perspective of concurrent engineering, especially Walter Sandig and Dr. Georg Haussler in the Nurnburg site, and to Elaine Seigal for help in editing and indexing the book.

Finally, many thanks to my family for emotional support during the editing and production of the book, including my wife Jackie and Mike, Gail, Nancy, and Jon. Also, to the many attendees of my seminars on concurrent engineering and quality methods, including the in-company presentations, who kept my interest and faith in concurrent engineering, I wish them success in implementing concurrent engineering tools and methods in their companies.

TRADEMARK LIST

Following is a list of trademarks and tradenames used in this book.

Hewlett Packard
Northern Telecom
Raychem
Ford Motor Company
Mitsubishi Heavy Industries
Pacific Bell
Rockwell Defense Electronics
Chipcom Corporation
Chipcom ONline System
Post-it
Mercury Computer System Inc.
Sun Microsystems Computer Corp.
Bell Canada Enterprises
Bell Northern Research
Polaroid Corporation
IBM
Sony
Digital Equipment Corporation
Bull Worldwide Information Systems, Inc.
Greene-Shaw Electronics
Zenith Data Systems
General Motors
Texas Instruments
Motorola
AT&T
Raytheon
Sanders
MA/COM Ominispectra

Successful Implementation of Concurrent Engineering Products and Processes

1
Concurrent Engineering in New Products and Processes

Sammy Shina
University of Massachusetts, Lowell

Concurrent engineering is becoming a standard model for new products being created at companies throughout the world. It is a methodology for creating timely products while maintaining the highest quality, lowest cost, and most customer satisfaction. Concurrent engineering is being practiced in both large and small companies, for creating many different types of products.

NEW PRODUCT TYPES

There are three types of new products being created in today's global competitive environment:

1. *Breakthrough products*—These are technology-based products, which are driven by basic research in the national or major corporate laboratories in new materials or technologies. The technology spawns product ideas, which are brought over to the development organization in companies through a process of *advance development,* where product feasibility is determined at both the design and manufacturing stages, in a timeline that is usually 3–4 years before actual commercial production. This phase is outside of the tight, new product development schedule (12–18 months), but has a definite set of requirements, budgets, and timing goals to be met. The products are then scheduled for *development* through the product/project management process.
2. *Follow-on products*—These are new products within existing product families that are introduced by industrial companies, depending on the price/performance characteristic of the marketplace. They are driven by changes in customer expectation, competition, technology, market devel-

opment, and opportunity. These products are defined internally by a combination of inputs from sales, product management, and the line managers. They normally go through a regular development schedule, having minimum interface with advance development or research labs.

3. *Modified products*—These are enhancements on existing products that are made essentially for cost reduction or specific customer inputs or options. They are very simple in nature and require little engineering effort. The modified products are usually handled by specially assigned engineers in the production department. They sometimes require a thorough knowledge of the manufacturing process as well as the technology of the product and the material choices involved. In addition, there is the advantage of shielding the development engineers from current production problems and modified products. It has been estimated in some U.S. companies that as much as 50% of development engineering time is spent on handling current production problems. This could adversely affect the new product schedule.

Of the three types of products, the breakthrough products have the largest impact on the long-term financial health of the company, as well as the greatest risk. The experience with breakthrough products at Hewlett Packard is typical. In the 1970s the hand-held calculator business was one-third of the company's sales and profits, which disappeared quickly as the competition reacted. In the 1980s the laser printer product family, which was originally developed by three engineers, made more money for the company than the products of thousands of engineers in the computer business.

In summary, the following compares the two major product types:

Breakthrough Products	*Follow-on Products*
Complete development process (Research-Advances development)	Mostly development phase required
Long total development cycle (3–4 years)	Short cycle (3–18 months)
Large impact on profits (home run)	Competitive positioning and market share protection (cash cow)
Technology and innovation driven	Cost and market driven
Thrives in uncoordinated processes (skunkworks)	Requires a systemic process
Many competing designs (high development risks)	Dominant or proven design (risks are smaller)

A good company product strategy is critical to its long-term survival. The correct mix of product types and the timing of product introductions depend on many factors, including the maturity of the marketplace, the state of the competition, customer trends, and the state of the art and technology.

Clearly, the impact of new products is very critical, as indicated by typical vintage charts at different companies. In Northern Telecom, 70% of the total revenues of the company come from products introduced during the last 4 years. The correct risk-versus-reward ratio in the new product creation process is what ultimately determines the future success of the company.

THE MATRIX ORGANIZATIONAL STRUCTURE

Product strategy will determine what the structure will be like that supports the product creation process. Historically, companies tended to organize in a matrix structure, with a central focus on traditional functions: engineering, marketing, and manufacturing. New products were based mostly on breakthrough ideas, and product introduction cycles were long because of the time needed to develop the technology and make it manufacturable.

In a matrix organization, a central engineering department plays an important role in developing new products. The department is broken down into subdepartments, which are organized along the traditional disciplines: electrical, mechanical, software, and support. New products are initiated by technology trends and run by project managers, either from the product management department in marketing or from engineering. The focus is on managing the schedule and the engineering and material resources. Engineers are borrowed from a central engineering pool, with a dual reporting scheme—one to the discipline or resource manager for technical guidance and the other to the project manager for work schedule. All communications, negotiations, and directions originate from the project and resource managers. Engineering is considered to be another commodity, like capital and equipment.

The matrix structure is clearly designed to maximize resources for *breakthrough products*. The matrix organization is normally focused on the process, not the product. The work is negotiated by the managers, and instructions are then handed down to the workers. Matrix management creates an environment that discourages engineers' creativity and leaves the managers with all of the decision-making powers. It is a system that discourages direct engineer-to-engineer communication, since interdisciplinary decisions have to be negotiated between the project managers and the resource managers. The system is supposed to control engineers' activities by micro-managing their time. Actually, it tends to shield the engineers from interfacing with many other activities in the company, such as manufacturing, marketing, purchasing, and production, and hence their designs tend to reflect their lack of understanding of how to maximize these resources for their product's benefit. In addition, longer communication lines produce misunderstandings and delays in conveying critical information, making for a poor and untimely decision-making process.

Matrix management has been very successful in large companies with vertical integration, diverse geographic resources, and corporate research laboratories in which to investigate and develop new technologies. Success depends on individual resource and project managers who negotiate resources, schedules, and product performance.

Problems arise when the interfaces are blurred between functional and resource departments; therefore, one-on-one negotiations play a greater goal in achieving common objectives. Different aspects of product creation (such as the division between research, advanced development, and development) are not specifically defined, and the formal transition points are mostly dependent on the managers and the circumstances involved. Therefore, the product creation process could be different in the same company, because it is shaped by historical precedence as well as by the personnel in the business unit.

Recently, the matrix organization and its long communication lines and process focus have tended to isolate the company and its management from its customers and market trends. These companies have been forced to restructure into more product-focused organizations.

THE PRODUCT CREATION PROCESS

Product creation is the correct mix between technology advancement, market conditions, consumer trends, and competitive factors. Planning is the key to developing a coherent product introduction stream that anticipates the market mix. Otherwise, product creation becomes a reactive process, with the subsequent risk of developing a product too late to capture a significant presence in the market or of having to lower the existing product prices to protect market position.

The worse scenario is the scheduling of a new product with unrealistic expectations or timeline. It places tremendous strains on the organization in general and the development team in particular. In addition, marketing plans, which are set in motion based on the false schedule, will be undermined for both existing and new products.

Product creation should be a team-centered activity. The balance of the mix between technology and market input can be determined best by practitioners of both crafts. The inputs from marketing and sales organizations, the research laboratories, the advance development group, and the current development organization should be evaluated, and a collective decision reached for the timing of the next product rollout. A risk benefit analysis should be made of the trade-offs between rushing a new technology or idea to market versus properly investigating the development and manufacturing problems.

Properly performing the traditional marketing functions in its entirety is critical.

These activities include the following:

Strategic market analysis
Competitive analysis
Product planning and positioning
Product business plans
Marketing communications
Market and customer surveys

PROJECT MANAGEMENT MODELS

The project management model is directly connected to the organizational structure. Matrix organization tends to operate in a *management-directed project management*. In this model, the emphasis is on the managers of the functions negotiating with each other the project requirements needed to meet the project goals. The project is *schedule driven,* and the decisions are made by managers and acted through instructions to workers. Alternatives are left to the responsible manager to make a yes or no decision on meeting the project milestones.

Concurrent engineering is best suited for *team-directed project management*. In this model, the emphasis is on collective decision making. There is a core project team that manages all project activities and that is supported by sub-teams of the different functions involved. The team members are the individual workers who are performing the specified tasks. The project manager is the *driver* of the project, making sure the schedule is on time and within budget. There is a strong technical component to this project management, and decisions are made by the project manager, based on both technical and business evaluation of the project conditions. Trade-offs are discussed collectively and decided upon to maintain the overall goals. These trade-offs include project scope, resources, and schedule. The differences between the two types of project management are as follows:

Management Directed	*Team Directed*
Project manager as *scheduler*	Project manager as *product champion*
Workers operate on many projects simultaneously	Workers assigned to one project

Tasks negotiated by managers	Team has a common vision
Long communication loops between workers	Direct communication loops
Managers assume and delegate responsibility	Workers and managers agree on commitment
Workers are shielded from other functions	Workers are exposed to the total organization

AXIOMS FOR CREATING CONCURRENT ENGINEERING WITHIN THE ORGANIZATION

In order to create a concurrent engineering environment within the company, there should a move *away* from the process-based organization, with resource (matrix) management reinforcing the use of a standard model for efficient and specialized use of resources. This model is a good one for managing a large and complex set of products and systems. In the current conditions of worldwide competition, the smaller, more efficient organizations are the ones that are nimble and can react quickly to the market. They are successful because of their focus on their business unit products, and they can make fast decisions because they are micro-managing a small organization instead of a large business entity.

The axioms for creating a new product creation process are as follows:

1. Create a *total quality culture* within the organization.
2. Introduce a *product focus* organization at the business unit level.
3. Emphasize the *team focus* approach to project management.

The implementation of these axioms at each functional level could be as follows.

Create a Total Quality Culture within the Organization

Total quality is a critical element of success for concurrent engineering. It is the base from which all other ideas, procedures, methodologies, and tools of concurrent engineering can be developed, nurtured, and successfully implemented.

Although the focus of a total quality culture is to control and enhance quality, it has evolved into many highly successful methodologies for many different aspects of successful management and operation of companies. Total quality infuses the whole organization with a common set of terminology and procedures to perform the following important tasks:

1. *Problem identification and resolution*—The organization is trained to spot problems (quality or otherwise), identify them properly, and suggest methods for improvement. Alternatives are studied and weighed carefully, decisions properly made, and adverse consequences evaluated. Management is kept informed and provides guidance, encouragement, and resources for successful completion of the tasks.
2. *Team process*—Total quality is synonymous with the team process. It encourages working in groups, helping team members reconcile individual versus group goals, set team objectives and expectations, make collective decisions, and learn to operate with less management direction. All of these elements will be very important for the successful implementation of concurrent engineering projects.
3. *Continuous improvements*—This is the idea of not being satisfied with the status quo, *not doing things the same old way,* and constantly seeking better performance from people, equipment, and processes. Part of continuous improvement is the challenge to realize that a limit has been reached with the current situation and to seek other alternatives and original ideas to improve. Expectations should be set correctly, and improvements can be achieved in *big* steps only when using new technologies or methodologies. Total quality allows for many *small* steps, which, when accumulated over time, lead to large steps.

One of the inhibitors of total quality at the engineering level is the view that it is for manufacturing and lower-skilled personnel in the company—"4th grade stuff." In addition, engineers by nature have been trained in universities to compete instead of collaborate: Grades and exams stress individual contribution rather than teamwork. These cultural inhibitors have to be dealt with by treating them as procedural issues as well as training issues. Emphasizing quality and teamwork on performance evaluations sends a strong message that the company is serious about implementing a total quality program.

The formation of total quality teams that provide for completion of successful projects at all the engineering and marketing functional levels is a good precursor and training ground towards a thriving concurrent engineering culture in the company.

Introduce a Product Focus Organization at the Business Unit Level

The company or the business units of a major corporation are comprised of a collection of functions required to manage a corporation, including the traditional three: marketing, development (or R&D), and manufacturing. In the following text, each functional unit is analyzed and issues are outlined for successful support to the concurrent engineering process.

Marketing/Product Management

The marketing department is normally responsible for formulating plans and strategies to penetrate existing markets and open new ones profitably. The marketing department implements programs to support the field sales force. Market research and analysis is conducted to provide R&D with information on new market opportunities.

Marketing identifies trends and forecasts general business activities, providing long- and short-term forecasts. Detailed product sales forecasts are provided to establish production schedules. The department evaluates and reports on competition, providing information on trends, market share, and product features and positioning. It directs, in cooperation with the sales force, a continuing and coordinated program for sales and promotion, including brochures, specification sheets, and so on. It recommends new product pricing and performance levels and implements the introduction of new products into the field.

Product management is the marketing representative on the new product project team. The product manager provides the prime interface, from the product to the market, while the project manager should provide the primary interface with the technology, according to the schedule time and cost goals.

Development is responsible for the design of the product and, jointly with product management, evaluates the design, making sure that it meets the market needs and justifies the project investments.

The formal processes for these important marketing functions should include the following:

Marketing research and analysis (market segmentation)
Product definition and positioning
New product business plans
Business development

A *business development plan* could consist of the following:

1. Statement of the purpose of the plan.
2. Specific market objectives to be achieved within the intermediate period (3–5 years).
3. Description of market, including potential customers and channels of distribution.
4. Description of the competition, their technical capabilities and potential, and their current product profile and emphasis.

5. Description of products and services necessary for success in the next 3–5 years, and the plan for development or purchase of such products and service plans.
6. Financial analysis of costs and returns.
7. Risk assessment.
8. Tactical implementation plan matching the tactical business horizon (12–18 months).

The Development (R&D Engineering) Department
The development department is responsible for the following functions:

1. A *communication function* with the corporate research laboratories or the general technology base of the company's business for initiating new product ideas and technologies.
2. An *advance development function* to transfer new technologies into product and process feasibility. It is beneficial to isolate this function from the tightly scheduled development effort.
3. A *development function* with cost and schedule control under project management to create new products.
4. A *technical support function* for the design systems that the engineers are using.

Technical problems in production should be handled by the manufacturing engineers. Otherwise, it can be a hidden impediment to timely completion of new product schedules. In some recorded instances, as much as 50% of the development engineers' time was consumed by solving technical problems in production. The contrary argument that says that design engineers are responsible for solving production problems is resolved by concurrent engineering, since manufacturing engineers are on the new product team.

New product development projects are staffed by a team of multidisciplined engineers and managed by project managers, who could be management appointed or team selected from either the development ranks or product management in marketing. The development department manager is responsible for the technical content and the technology to be used in the development project. He or she interfaces directly with the project manager on solving technical problems, as well as coordinating the project schedule and the assignment for tasks.

The Production Department
The production department is responsible for the following functions for products and processes:

1. *Current products support,* including the scheduling, planning, documentation, assembly, and test. Production should be also be maintaining a continuous improvement posture to achieve higher quality, lower cost, and superior performance within the advertised specifications of the product.
2. *New products introduction,* including manufacturability assessment, production scheduling and layout, material acquisition plans, product data transfer, test development for inline as well as final test.
3. *Technical support* for automation processes, such as computer-aided manufacturing CAM and product data transfer.
4. *Process development* for assembly, test, and automation technology.
5. *Communication* with the other parts of the organization—new product project management, logistics, materials, and quality and development departments—to insure meeting shipments and targets.

The production department performs all the functions necessary to maintain a viable shipping profile. The interaction and communications with the logistics, materials, quality, and other support functions, as well as marketing and development, should be open and bidirectional.

EMPHASIZE THE TEAM FOCUS APPROACH TO PROJECT MANAGEMENT

The team concept for project management has been shown to be very effective for new product development. In small projects, the product manager can double as the project manager. In this case, the original product specifications can be preserved through the development implementation, by having one person responsible for both technical and project leadership.

In large projects, depending on the concurrent engineering maturity of the organization, either an appointed or a de facto development project manager emerges who has the final authority to resolve, with other team members, any technical problems that may arise. There is a core team that is composed of all the people who work on new development, from concept to production to field performance. Members of this core team are the prime experts (or initially managers) of the different functions involved.

Reporting to this core team are the members of each activity; they are identified by job function and contribution to the project tasks. There are separate teams for electrical, mechanical, software, production, logistics, and other functions, grouped by skill sets, who report on their progress to the core team. In this model, cross-functional communications occur directly from one team member to another, without having to deal through a management function. *Cross-functional teams* should be encouraged, since they shorten communication loops.

The challenges to successful concurrent project management are as follows:

1. Poor visibility of external and internal dependencies and their potential impact on the project schedule.
2. Focus on the technical specification and lack of coupling to financial, logistics, quality, production, and other functional expectations.
3. Lack of standard communication vehicles, both up to management and down to individual workers.
4. Clear understanding of management goals and expectations.
5. Strong belief in the benefits of planning and executing project management guidelines and techniques.
6. Difficulty in allocating people in a dynamic environment and in defining the skill level necessary to successfully complete assigned tasks.
7. Unclear definition of project objectives and specifications and of reactive market decisions during the development phase.

The issues when implementing concurrent projects are:

- For management, it is the feeling that the project is on schedule, within the prescribed cost and with the defined functionally. *This feeling* **intensifies** *as the project nears completion.*
- For the development engineer, it is the clear understanding of the tasks required, the importance of the task in the overall project plan, and the amount of risks involved in meeting the technical objectives. *When to report that a* **schedule buster** *problem has occurred?*
- For the project manager, *how to operate between the rock and the hard place.*

Phase Review and Concurrent Project Management Control

One of the proven methods for alleviating unsuccessful project management is the use of the *phased review* technique. Specific phases are identified. Each phase can be viewed as a stand-alone entity with objectives, deliverables, product cost, quality, serviceability and manufacturability status, and project costs to date. The phase review process brings together the core project team and a selected management group at the end of each phase (milestone) to review the status in terms of achieving objectives, analyze recommendations, make appropriate decisions, and commit to the next phase.

The spacing of these milestones is very important. They should be scheduled as required in groupings of like tasks, with more and closer milestones as the project nears completion. The functional and top management should use the milestone meetings as an opportunity to review in detail the project and its current direction and fit to the changing overall objectives.

Each functional area should have specific plans, measures, and goals as part of the overall project plans. These plans should be reviewed at the milestone meetings in addition to the technical review of the project. A sample of the plans could be:

Manufacturing should have testability, quality, yield, and throughput goals. Responsibilities include producibility feedback, test strategy and plans, production documentation, technical competence to handle the product after release to production, operator training, review of the manuals, and updates on production equipment installation and production process optimization.

Product management should review the competition, market surveys, profitability, pricing, obsolescence, and overall scheduling at each milestone meeting with the general management.

Sales should review the forecast, the product introduction plans, the promotional plans, and the feedback from customers and dealers. *Service* should review the mean time between failure (MTBF) and mean time to repair (MTTR) and the service and repair plans, resources, training, and equipment.

Quality should review the product quality specification according to the Quality Milestone plans.

Materials should review part procurements and qualifications. Suppliers status, especially overseas, should be also reviewed.

Controller department should review expenditures to date, as well as remaining funds and timing of major purchases. In addition, the costs and profit calculation should be redone if there are any changes.

A good project management plan will contain the schedule details, as well as input from all functions necessary for the overall success of the product. A set of goals for the plans should be:

1. Provide concise project definitions with phases and milestones, with specific deliverables at each phase.
2. Identify team members and their responsibilities, and keep management and team members informed of progress.
3. Force team members and managers to continuously evaluate and replan.

4. Spot potential problems, and help in taking preventative action in time.
5. Improve communications and delegation of duties and responsibilities.
6. Help the team focus on the activities at hand.

Key Processes to Enhance the Concurrent Product Creation Process

These processes can be developed to operate in the suggested organizational structure in order to create an environment for continuous improvement and operational excellence through concurrent engineering:

1. Phased review process—This should be the primary vehicle for project management. Specific phases are identified in the product development process, with each phase being a collection of task completions. Each phase can be viewed as a stand-alone entity with objectives, deliverables, product cost, quality, serviceability and manufacturability status, and project costs to date.

    The project team should plan each phase and milestone carefully, with shorter time between the later milestones. The process should be used as the primary vehicle to update the management and project teams with the current status of the project. The phase review process brings together the core project team and a selected management group at the end of each phase (milestone) to review the status in terms of achieving objectives, analyze recommendations, make appropriate decisions, and commit to the next phase.

2. Quality advocacy and the quality systems review—This procedure is used to assure that the quality system is effective in achieving Total Quality and customer satisfaction. The historical focus on regulatory and product quality and reliability issues should be augmented by a quality advocacy at each functional level. The quality organization at the company's highest management level should put sufficient emphasis on facilitating an company-wide adoption of total quality methods (TQM). It should have a process rather than a product focus. The organization should have a consulting role and is assigned to assist other parts of the company in integrating quality methods in their day-to-day operation.

    The quality systems review is an assessment vehicle that evaluates the status of quality in each function and department. It defines the *quality vision* of how business should be conducted; it sets a common goal of quality and provides an awareness of quality systems requirements across the organization. The quality system review process should be used as a measure of the quality progress, to provide the opportunity for exchanging ideas and to serve to refocus each part of the organization on the basic issues of quality.

3. Manufacturability assessment—This process evaluates new products for ease of manufacturing, to ensure a high level of quality, and to maintain lower production costs. The issues of production ease can be outlined in the following areas:

   a. PCB assembly—this assessment ensures a high degree of automation in the assembly areas, by ensuring that only approved parts are used, in the quantity and variety that can be completely handled by the existing production equipment.
   b. Testability assessment—This operation will ensure that all tests, including in-circuit, final, and systems tests can be handled adequately, by ensuring proper physical access to the PCBs and the product, using appropriate test methodologies, and integrating production and other service and self-test procedures and algorithms.
   c. Product assembly—This assessment ensures that proper automation, both current and future, can be applied by reducing the number of distinct parts, assembly motions, part geometry and symmetry, and other aides to enhancing automatic and robotic assembly.

Tools to Support Suggested Processes

A list of tools is available to support these suggested processes. Training on these tools should be provided to assist in achieving success in applying these tools.

Tool	*Focus*
Entity strategy	Goal setting at each level
	Set "competitive advantage"
Engineering process	Engineering process metrics
	Project management/review
	Cost estimating tools
Structured analysis	Document MFG Process
	Design new system/process
Process capability	Set each MFG step rejects
	Negotiate with suppliers
	Final product yield
Control charts	Monitoring MFG Quality
	Input into process capability
DFM analysis	Axiomatic theory
	Conceptual design
	Assembly efficiency ratings

Design analysis   Robust design
                       Design of experiments
                       Reliability analysis
QFD   Customer focus engineering
                       Marketing and engineering

A detailed discussion of these tools is available in a book authored by the editor, Sammy G. Shina, entitled *Concurrent Engineering and Design for Manufacture of Electronic Products,* published by Van Nostrand Reinhold of New York in 1991.

2

QFD Comes to Raychem Corporation: The Story of a Pilot Project

Marilyn Liner
Raychem Corporation

CHAPTER OVERVIEW

Incorporating *quality function deployment* (QFD) principles and methodology into the new product development process offers important benefits in bringing new products to market efficiently and increasing customer satisfaction. This chapter describes a product development team's first experience using QFD: the charts used, the benefits and difficulties experienced, and the evaluation of the process by the team members. The matrices used are related to both of the principal QFD approaches taught in the United States today: the Four-phase approach [2] and the Matrix of Matrices approach [7]. A special matrix developed by the team to answer a specific question raised during the development work is also described. As a result of the successes achieved using QFD methods in this and other projects across the company, QFD elements were incorporated into the company's process for new product development.

COMPANY PRODUCTS AND STRUCTURE

Raychem Corporation develops and manufactures a wide variety of industrial products utilizing materials science technology. Raychem's technologies include high-performance cross-linked and conductive polymers, elastic memory materials, adhesives, gels, ceramics, shape memory alloys, thin films, liquid crystals, and advanced materials engineering.

The company's products include environmentally sealed connectors and connection systems for telecommunications and power transmission applications, self-regulating heating cables for process industries, lightweight wiring harnesses and interconnection systems for aircraft and automotive applications, and specialized components and connectors for the electronics industry.

A Fortune 500 company with offices in over 40 countries, Raychem has approximately $1.2 billion in sales and employs over 10,000 people worldwide. Raychem's seven primary independent business units belong to three business sectors: electronics, telecommunications, and industrial. Each business unit focuses on particular products and product lines, and each has its own general management, manufacturing, marketing, development, and sales organizations. Central corporate groups provide support to the product divisions in areas such as human resources, facilities, communications, finance, and advanced development for materials, products, and processes.

Product development responsibilities are shared by the product divisions and the Corporate Technology group. Development engineers in the business units concentrate upon new product applications utilizing existing technology, while the corporate group provides technical and creative expertise in identifying and applying new technologies. Corporate Technology also provides support to the business units' new product development efforts in the form of specialized services.

Context and Company Culture

Raychem Corporation is regarded by its customers as having excellent product quality. Each year, the company receives quality awards from customers. In recent years, Raychem business units have received supplier excellence awards from companies including Ford Motor Company, Mitsubishi Heavy Industries, Pacific Bell, and Rockwell Defense Electronics.

Yet despite these honors, increasing competitive pressures and changing market conditions have resulted in a shift in focus within Raychem toward improving the efficiency and effectiveness of the internal company processes used to bring products and services to customers—from financial procedures to manufacturing, order entry, and the product development process. The increasing emphasis on improving company internal processes by employing the philosophy and tools of total quality management (TQM) has created changes within Raychem. Tools such as statistical process control, design of experiments, and QFD are now regarded as essential. The prevalence of self-directed work teams is increasing, and factory organization around manufacturing cells has become the norm.

The emphasis on TQM is creating changes in the traditional company culture. For example, because individual resourcefulness and independent action had been highly rewarded at Raychem in the past, employees were not accustomed to working in a team environment or to using structured methods. In contrast, TQM demands that team members from different functional departments collaborate using shared approaches and tools to standardize and improve work processes.

QFD, one of many tools supporting the development of a TQM environment, is a structured methodology serving the product development process. QFD's promise for product development—shorter time to market, lower development costs, and delighted customers—has increased senior management's dedication to leading the cultural changes needed for continuing business success.

Product Division Background

This is the story of the first new product development project using QFD in Raychem's Telecom Division in Menlo Park, California. The Telecom Division is a $275 million division providing products for the telecommunications industry. These products include environmentally sealed telephone line splice cases and connectors, fiber optic splice cases and other fiber optic equipment, and coaxial cable connectors.

QFD's structured approach to concurrent engineering was a radical departure from past practices. Product development had typically proceeded in a sequential manner—engineers did not directly gather customer data, and manufacturing involvement did not occur until late in the development cycle. Obtaining input from customers usually presented difficulties. Customers were often asked to respond to new technical solutions with which they had no prior experience and that had not been specifically developed to address their particular needs. In the past, emphasis on speed in reaching the market with a new product had often been followed by intensive and costly efforts to make up for inaccurate assumptions about product requirements.

Telecom management was particularly interested in QFD as a possible way to speed up the development process. QFD fit in well with efforts underway within the division to improve project management techniques, emphasize market-driven opportunities, increase engineer/customer interaction, and operate using multidisciplinary teams [11].

SELECTING THE PILOT PROJECT

Starting with introductory information about QFD that was received through the Raychem Corporate Quality group, the Telecom management team decided to pursue further knowledge by sending a cross-functional group of managers to a 2-day overview seminar on QFD. Managers from the marketing/sales, manufacturing, and development groups attended. The decision to use QFD in a pilot project was made shortly thereafter, championed by the technical director. The goal of management in launching the pilot project was to see how QFD worked and to evaluate its use in future projects.

The project chosen was an indoor connector for the cable television (CATV) market. A relatively simple product compared to others developed by the Telecom Division, the indoor connector offered the chance to experiment with QFD and to see the results within a relatively short time period.

The transfer of initial Corporate Technology feasibility work on the connector to the Telecom Division for testing and commercialization was imminent. Telecom management realized that the maximum benefit from using QFD would be gained by applying the tool early in a project, before selection of a specific design solution. However, the indoor connector project presented an attractive opportunity for Telecom to use QFD in facilitating the transfer process, and the project was selected.

The development of the indoor connector was a strategic move by the CATV Group within the Telecom Division to originate an important new product line. As a result of company experience with supplying premium environmentally sealed outdoor connectors to the CATV marketplace, the idea of segmenting the market into indoor and outdoor application groups had emerged. A high-performance CATV connector designed specifically for indoor use was not available to North American cable system operators. This situation laid the groundwork for the decision to differentiate and produce a product to meet the specific needs of the indoor market segment.

Customers, Marketplace, and Product Application

Thorough knowledge of the customers and marketplace is an important success ingredient for a product development effort, with or without QFD. The Telecom Division enjoyed close working relationships with many important customers because it already had products in the CATV marketplace. The customers for the new connector—approximately 100 system operators nationwide running some 8,000 cable systems—utilized tens of millions of connectors on indoor equipment annually.

The CATV connector team had a clear definition of the specific application environment and the benefits to customers that a successful new product would provide. The indoor connector (see Figure 2-1) was a push-on connector for CATV operating companies to use on home video, television, and cable converter equipment. The indoor connector saved time for cable operators at installation and allowed homeowners to easily disconnect and reconnect their own equipment.

These advantages meant fewer trouble calls to the cable companies from

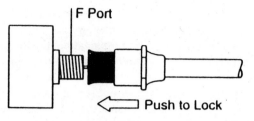

FIGURE 2-1. Sketch of CATV connector.

subscribers and fewer service calls ("truck rolls") by CATV system operator repair staff. The indoor connector offered customers push-on convenience with a high level of technical performance—a combination of benefits previously unavailable in that marketplace. Existing push-on connectors were avoided by CATV operating companies because they were easily distorted by physical abuse, resulting in poor radio frequency (RF) shielding and poor picture quality.

Project Context

Work on the indoor connector design had first begun within the Corporate Technology group over a year before the start of the QFD study, with the understanding that the preliminary conceptual work would be turned over to the Telecom Division for commercialization. Using existing knowledge of the product application environment, two engineers in the corporate group had developed a unique, patentable connector design. The design solution for the indoor connector was completely different from that of the company's existing outdoor CATV connectors, where the need for environmental sealing was a driving factor.

When the CATV connector QFD study was started, relatively few design changes could be made, since product development work had reached the prototype stage. Yet, despite this disadvantage, both the CATV connector team members and management believe that important benefits were gained and that the use of QFD significantly improved the product. The team recommended strongly, however, that future projects apply QFD from the very beginning.

Forming the Team

At the direction of the Telecom Division senior management, a nine-engineer cross-functional development team was formed. The team included both Telecom and Corporate Technology members. Three Telecom development engineers joined the two original design engineers from Corporate Technology, along with Telecom engineers representing the marketing, sales, manufacturing, and quality functions. Purchasing contributed to the team later in development work, on an ad hoc basis, to assist with vendor assessment and selection. A Telecom development engineer acted as project leader.

Prior to the startup of the Telecom development team, the project leader and three other Telecom team members had attended training in a four-phase approach to using QFD [2]. The quality engineer, whose background included product development experience in marketing, manufacturing, and design, as well as an academic background in product design and development, had received training in both the Four-phase and the Matrix of Matrices approaches to QFD. The quality engineer served as QFD facilitator, guiding the team in the process.

Those team members who had not attended training before joining the team learned to use QFD through direct participation in the process of data gathering, chart building, and analysis while working on the project. Before starting each step in the QFD process, those who had attended training classes discussed and reviewed their understanding of the specific objectives, tools, and procedures associated with that step. This approach served simultaneously to align the QFD-trained team members and to teach the others about the process.

Starting the QFD Study

The CATV connector project was at a difficult stage when the cross-functional development team began its work using QFD. Although Corporate Technology was ready to transfer a prototype connector to the Telecom Division for detailing, manufacturing, and marketing, Telecom development engineers believed that design changes were needed. In addition, Telecom manufacturing was alarmed by the low cost targets. There was poor communication on key issues, and working relationships did not encourage open questioning.

Fortunately, as anticipated by Telecom management, using QFD in the CATV indoor connector project provided a structure that facilitated intergroup communication despite these initial difficulties. The fact that improved communication was experienced with the first use of QFD convinced Telecom management that QFD could be a useful tool for teams working to reconcile differing organizational perspectives within Raychem's highly matrixed environment.

The indoor connector team started by formulating a statement of its charter. The team's charter statement was: "Develop a CATV connection system for indoor use in the U.S. market, to fit all U.S. cable and port sizes, for installation by cable operators." This exercise was valuable, as it enabled the team members to resolve some initial disagreements about the markets and applications for the connector.

DEFINING CUSTOMER REQUIREMENTS

The first phase of QFD work involves identifying the customers and then gathering and organizing data to define their needs and requirements. This process is known as listening to the Voice of the Customer [2, 12]. While product development team members may often have a fairly good idea of what customers need, the QFD approach teaches that there is no substitute for gathering fresh data and setting product development priorities based upon facts rather than opinions.

The CATV connector team realized that the QFD process of listening to and responding to the Voice of the Customer is not the same as holding a meeting to brainstorm about what customers want. However, the group needed a starting point. Company beliefs about customer needs were written down, one per 3" × 5" self-sticking note card. A team meeting was spent grouping the cards into related clusters, using the affinity diagram technique [4]. After the meeting, a tree diagram [4] was drafted by one member for review by the team. A tree diagram breaks down a topic into successively refined levels of detail, and may be compared to an organizational chart placed on its side. Figure 2-2 shows an excerpt from the tree diagram of customer needs generated by the team.

Gathering Data from Customers

To get actual data on customer needs, interviews were then conducted with customers. The initial tree diagram served as a guide for drawing out customer discussion via open-ended questions. The sales and marketing team

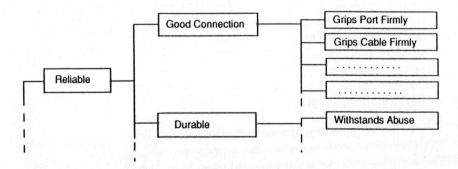

FIGURE 2-2. Tree diagram of customer needs.

members conducted telephone interviews with two dozen familiar customers representing different-sized CATV operations and different geographical areas. Direct quotes from the customers were written down, along with their sources. Due to the strong personal relationships that already existed with contacts at the customer companies, the telephone approach yielded information quickly.

In evaluating the process used to collect the initial customer data, the team recommended that, in the future, all team members have the opportunity to meet and talk directly with customers early in a project. In the CATV connector project, many team members did not get to meet customers until the field trial stage, some 6 months after the first customer telephone interviews were held. The team members also recommended that future teams use an outline of consistently phrased visit questions to gather customer data scientifically. Because not all topics had been covered with each customer during the initial interviews, repeat telephone calls were needed to fill in information gaps.

Now, 3 years after the first team's experience, customer visits by product development team members are standard procedure within the Telecom Division for new product development projects [9]. Teams routinely plan discussion guides to assure gathering consistent data, and schedule opportunities for members to visit customers and observe customer work environments early in development projects. Team members meet with persons in a variety of functions within customer companies, and work to understand the decision-making processes within the companies and marketplaces that their products will serve.

Analyzing the Customer Voice

For the CATV connector team, the next task was to analyze the relationship between the customer responses and the group's initial list of customer needs. The question needing resolution was, "What is the relationship between what the customers said and what we originally thought?" The answer to this question had to be found before the team could progress to the first traditional QFD matrix, which requires an accurate list of customer needs as a starting point.

As the team discussed how to answer this question, it became apparent that the quotations gathered from customers often related to several items on the team's initial list; a one-to-one correspondence did not exist between the items within the two sets of information. The analysis would also need to account for the particular importance of certain customers who had provided data and would need to reflect the fact that some items had received attention from a large majority of the customers interviewed.

As this discussion progressed, the team developed the idea of creating a special matrix to analyze the information. The matrix tool appeared extremely appropriate for this work, as it would array the two lists of items against each other and would allow for weighted multiple relationships to be displayed and summarized.

While QFD is usually presented as a series of four or more standard matrices [2, 12], in actual practice it is frequently valuable to use a nonstandard matrix format to answer specific questions relevant to the project at hand. These questions will be somewhat different for every project.

An important activity of teams using the QFD approach is to construct matrices displaying "whats" versus "hows." These matrices assist in making decisions and in determining where resources should be spent in product development work. QFD is thus a flexible tool that can be used creatively to assure that a company focuses its primary efforts on providing those product or service features that are most important in achieving customer satisfaction.

With the goal of applying the QFD principle of matrix analysis to the problem of relating the initial internal list of customer needs to the actual customer data, the team constructed a matrix displaying customer quotations ("whats") versus the preliminary internal list of needs ("hows"). An excerpt from this matrix appears in Figure 2-3.

Each customer statement was assigned strong, medium, or weak relationships to the items on the initial list. A weighting factor for each statement

Relationships: ■ Strong = 9 pts, ▤ Medium = 3 pts, ☐ Weak = 1 pt CUSTOMER QUOTATIONS	LIST OF CUSTOMER NEEDS (Initial Assumptions) →			Easy Instructions	Installs Right the First Time	One Handed Installation	Easy to See in the Dark	
	Quotation Weighting**							
•	•				■			☐
•	•	▤	■			▤		
•	•			■				
"Must be easy to re-enter"	•							■
"High degree of RF shielding integrity"	•	■						
"Worried about damage to ports"	•							▤
"Secure, will not back off over time"	•							
•	•			☐	■			
•	•					▤	☐	
•	•				☐			
Weighted Customer Needs		•	•	•	•	•	0	•

** The weighting factor for each customer statement tallied the number of customers mentioning the item. The more important and technically knowledgeable customers were counted as 1.5, others as 1.0.

FIGURE 2-3. Customer voices vs. needs list.

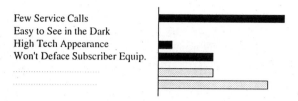

FIGURE 2-4. Summary of weighted customer votes.

reflected the strength of the customers' opinions and frequency of mentioning, as well as the technical knowledge and market influence of the source. Statements made by customers who had more technical knowledge or were more influential in the marketplace received higher weighting factors. For each customer need item on the original list, an importance score was then calculated by summing the products of the relationship values and corresponding weighting factors. This standard calculation is used throughout QFD matrix work as a prioritizing technique [2, 12].

Individuals and subgroups of the team completed separate sections of this matrix manually, using pencil and paper. The marketing team member summarized the results in bar graph form using a personal computer spreadsheet program. A few entries from this summary appear in Figure 2-4.

Discoveries

Analysis of the customer data revealed that certain preliminary assumptions about customer needs were in fact wrong! For example, it had been assumed that the connector had to be visible in the dark. The interview data revealed that this attribute was not important to customers. As a result, fewer constraints were placed upon the design as well as on the material choice. This data point was later reconfirmed when team members accompanied CATV personnel on installation visits during product field trials.

The survey process also revealed that customers did not emphasize several essential functional needs. For example, the need for compatibility with commonly used ports and cable sizes was an absolute requirement, yet it received a low score during the customer interviews. The tendency of customers to perceive and communicate their needs incompletely is explained by the Kano model [1, 2], which describes different types of customer perceptions. Often, as this model states, the most important functional requirements will be unspoken ones. This fact underscores the need for team members to visit and observe customers and their environments in person.

Before reaching conclusions about customer requirements, the team discussed and carefully analyzed the weighted scores resulting from the matrix calculations; some items were retained on the list despite low scores. As a last step, an importance weighting between 1 and 5 (1 = low; 5 = high) was assigned to each item on the revised list of customer needs. These weighting factors indicated the relative importance of each need item from the customers' viewpoint.

In retrospect, the CATV team members felt that constructing this particular matrix was unnecessarily time-consuming. They felt they could have reached the same conclusions more rapidly by visiting customers in person and deriving a final list of customer needs directly from discussions and observations. The team recommended strongly that future team members get out of the office and visit customers personally, rather than relying on in-house analysis of data gathered by others.

Creating an original matrix to answer a specific question demonstrates the power and flexibility of the QFD tool. Other Raychem teams have also created nonstandard matrices to explore specific issues—ranging from describing market segment characteristics to displaying fixed or flexible product/process design factors.

DEFINING PRODUCT TECHNICAL REQUIREMENTS

The team's refined list of customer need items and the relative importance of each need item to customers were now entered into the "Whats" area of a QFD product planning matrix [2, 12]. A portion of this matrix, also known as the House of Quality [6] or the A-1 matrix [7], appears in Figure 2-5.

Work then began on the "Hows" axis of the matrix. "Hows" are technical performance requirements for the product that translate the customer requirements, or "Whats," into internal company measures and specifications. The technical requirements are quantitative, design-independent product performance parameters that serve as predictors for achieving customer satisfaction with respect to the qualitative customer needs or benefits.

Work to determine the technical requirements [12] proceeded based on a combination of internal specifications, the existing technical product knowledge base, and team brainstorming sessions. The team adopted several different approaches. Areas closely connected to qualification testing, such as function/reliability and installation, were worked on as a group. Categories of lesser importance were assigned to individuals for later review by the team.

Between meetings, team members worked individually on tables and tree diagrams to translate the customer needs into technical requirements and to record written definitions for each technical requirement. Tables documenting the definition of each technical requirement, its related tests, and test value requirements were also generated.

QFD Comes to Raychem Corporation: The Story of a Pilot Project

		TARGET VALUES →	• •	• •	1 mode	xx lbs, x in-lbs	xx dB	x steps	• •
Relationships ■ Strong = 9 pts ■ Medium = 3 pts ☐ Weak = 1 pt		TECHNICAL REQ'TS →	• •	• •	No. of Installation Modes	Forces on Equip. Panel	RF Shielding	No. of Installation Steps	• •
CUSTOMER NEEDS		Importance to Customers							
•		•				■			☐
•		•		■					
•		•			■				
Clear Picture, Even After Abuse		•				■	■	■	■
Easy to Tell When Properly Installed		•		■		■		☐	
Long Lifetime		•							
Obvious/Simple to Install		•				■	■	■	■
•		•			☐				
•		•					■		
•		•							■
Weighted Technical Requirements			••	••	327	314	322	234	158

FIGURE 2-5. Customer needs vs. technical requirements.

At team meetings, copies of the individual members' drafts were handed out and discussed. The overhead projector provided a convenient way to record the technical requirements agreed upon by the team during meetings; writing on a transparency of a blank, preprinted matrix form worked well. Recording technical requirements using a marking pen on large wall chart matrices was also tried, but this approach lacked flexibility. During meetings, the team members generated additional technical requirements via discussion in order to ensure that all of the customer needs were completely addressed by appropriate product performance standards.

As a result of this work, the team members realized that they needed to talk further with customers to gain a better understanding of several items on the list of customer needs. The sales and marketing team members then got back on the telephone to obtain more precise information from customers. Because the CATV connector customers were technically oriented, they often described their needs in terms of technical specifications. These specifications were discussed in detail with the customers in order to allow the team members to fully understand the qualitative benefits lying behind each requirement.

QFD demands going through the process of separating qualitative "Whats" from quantitative "Hows" in order to correctly complete a product planning matrix. This process is very useful, because discussing with customers the reasoning behind specifications can lead to discovering areas of flexibility or new alternatives for delivering the desired benefits.

Managing the House of Quality Matrix

When the technical requirements for the indoor connector were complete, a working QFD matrix was prepared using a personal computer spreadsheet program plus photocopier "cut and paste." This first matrix was 28 × 52 items—too large to deal with easily. Overlapping levels of detail and names of tests that related to more than one performance parameter encumbered the first technical requirement list. Also, the list of customer requirements contained some items having very low importance to customers.

The matrix was revised to 19 × 13 items to make it more manageable. First, it included only those customer needs that scored 2 or above on the 1–5 scale of importance to customers described earlier. Then the technical requirements were regrouped for level of detail with the help of a tree diagram. For example, numerous entries detailing North American coaxial cable constructions and sizes were reduced to a single line item, "Fits all cable types and constructions." Instead of test names, technical requirements were substituted to reflect what the tests measured. An illustration of this reduction process using the tree diagram tool appears in Figure 2-6. The simplified matrix was documented using a personal computer drawing package and distributed to the team members.

At the conclusion of the project, the team members cautioned future teams against creating large matrices and recommended using fewer, broader customer needs and technical requirements. Because QFD work generates such a large amount of information, it is best to organize and initially prioritize the information before starting a matrix. A QFD matrix should contain just those items that the project team has chosen to concentrate on, not everything known about the product.

Establishing Target Values

The target values (see Figure 2-5) for each technical requirement came from a combination of discussions with customers, in-depth team discussions, and internal testing. For example, the range of push-on forces likely to be applied to the connector by cable subscribers was estimated using trials involving male and female Raychem employees. In other Raychem QFD work, designed experiments built upon a detailed existing technical knowledge base have been used to establish target values for technical requirements.

The CATV connector team decided to enter "world class" values into the matrix in order to reflect the industry technical performance levels required for market leadership in the next 3–5 years. These values were tabulated along with their corresponding qualification test methods.

Later in the project, it was necessary to determine the target performance values actually required for a successful product introduction. These targets

QFD Comes to Raychem Corporation: The Story of a Pilot Project

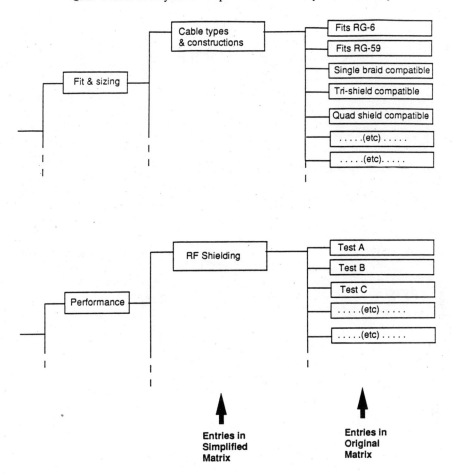

FIGURE 2-6. Use of tree diagram to manage matrix size.

would be reachable in the time allowed by the schedule. In retrospect, the team felt that time would have been saved if work to distinguish "world class" from required product introduction performance values had been completed sooner. However, the vision of world class performance determined by the team was useful in later product improvement work.

Prioritizing Technical Requirements and Identifying Trade-offs

To predict which technical requirements would have the greatest influence on meeting customer needs, the team evaluated the relationship between each customer need item and technical requirement as strong, medium, weak, or no effect. For each technical requirement, a weighted score was then calculated by summing the products of the relationship values and the corresponding importance factors of the related customer need items (see Figure 2-5). These calculations followed the standard practices used in QFD work [2, 12].

The team completed the relationships using the process of consensus [3]. This was not the same as voting or compromising. Where team members had differing opinions about the strength of a relationship, they discussed their reasons and came to a mutual agreement on the most appropriate value to enter into the matrix. Often, for example, one team member had knowledge or experience demonstrating that a particular relationship was strong, although other team members had initially judged it to be weak; the result of team discussion was to assign the strong relationship value.

The consensus matrix relationships were recorded during team meetings using an overhead projector. Members received photocopies of the official team matrix as it developed. Completion of the matrix, about 3 months into the QFD process, showed that certain technical requirements were most heavily weighted with respect to customer needs.

This information, along with the team's discussion and analysis, formed the basis for the product qualification testing plan. Testing was already in progress on RF shielding performance, an essential product requirement. However, more emphasis was now placed upon understanding other high-scoring technical requirements identified by the QFD process. Later in the project, field trials confirmed the central importance of the same technical requirements identified during the QFD analysis. This was taken by the team as a verification of the power of QFD.

The team discovered that matrix construction is something of an art. Some technical requirements received low scores but were essential to the product, such as the range of port sizes accommodated. The team concluded that, due to possible overlapping chart entries or mixed levels of detail, the best use of the matrix arithmetic results was as a guide for discussion, rather than as an absolute priority scale for ranking the importance of each technical requirement.

The lead project engineer completed a matrix showing potential interactions among the technical requirements that could lead to design compromises or trade-offs. This matrix is known as a correlation matrix [2] or A-3 matrix [7]. Although this work was not a team undertaking, it assisted the project leader in structuring the testing then in progress. At the conclusion of the project, the team members recommended that more attention be paid in the future to understanding such interactions and trade-offs.

ANALYZING THE COMPETITION

To assess customer satisfaction with each item on the list of customer needs, three competing connector products, an existing Raychem product, and the new concept were evaluated. The sales and marketing group members met to assess and tabulate the level of customer satisfaction with each of the identified customer needs. The group arrived at consensus scores based upon their discussions with customers during the initial telephone interviews. While the team had already discovered the benefits of clearly distinguishing between company beliefs and customer data, it was felt that communication with customers about the competing products was quite good. Feedback from customers on early prototypes provided information on the new CATV connector. Figure 2-7 illustrates the table and the 1–5 customer satisfaction scale used. This table commonly appears as a section within the House of Quality [2, 6] or A-1 matrix [7].

This point-by-point comparison among products on the market represented a new direction for the indoor connector development team; the structured discipline of taking a detailed look at the competition from the customers' perspective was a new process that had not existed in the company before the introduction of QFD.

The CATV connector team warns, however, that showing prototypes and announcing the product introduction date too early can encourage competition. Conscious business decisions must always be made to evaluate the relative advantages and disadvantages of exposing the details of a product idea to obtain customer input.

To further evaluate the existing competing connector designs, the team constructed an additional table comparing three competitors' designs with a

	Existing Connectors				New CATV Connector
	A	B	C	D	
Convenient Installation					
Customer Need A	3	3	4	4	4
Customer Need B	3	4	4	4	5
.... (etc). ...	3	2	4	4	3
Function/Reliability					
Customer Need N	4	4	2	1	5
.... (etc). ...	3	3	3	4	4

FIGURE 2-7. Analysis of the competition from the customers' perspective.

Raychem outdoor CATV connector and two new Raychem indoor connector design variations. Figure 2-8 illustrates a portion of this comparison table.

The comparison criteria were a combination of key customer needs, technical requirements, and sale price. Laboratory performance testing on competitive connectors as well as on Raychem designs provided data for the technical comparisons. This product comparison table resembled the Pugh concept selection table of Phase 2 [2], which is also known as the E-4 matrix [7]. The technical product performance comparisons recorded in this table are often documented in the A-1 or House of Quality matrix itself [12].

The indoor connector team's product comparison table clearly showed that the first Raychem design was superior to competing connectors in many areas, but still needed further improvement: it was more expensive, while offering no added RF shielding performance. Accordingly, work was redoubled to reduce manufacturing costs and improve RF shielding. With specific knowledge about what to improve and why, the design engineers continued to work on improving the connector. Reduced cost and higher technical performance were in fact achieved through design modifications in an improved Raychem design.

REFINING THE DESIGN

Throughout work to understand customer needs, the Corporate Technology engineers had continued to refine the connector design, incorporating the new input from the QFD process. Further input came from the Telecom development team members themselves. The Corporate Technology team members hosted a special laboratory session where the team practiced installing various connector designs on a variety of ports. Selected customers also received prototypes and then supplied feedback. The data gathered during this work led to additional design improvements that provided easier attachment to the coaxial cable, as well as better technical performance.

In the seventh month after beginning work with QFD, field trials with six widely dispersed customers began. Members of the team accompanied cable installers to observe and record their reactions to the new connector. At the field trial sites, a short questionnaire developed by the marketing team members helped to get direct feedback from the installers. Further design refinements were made as a result of the field trials.

As a status check, the team set up a written survey among its members to determine the principal remaining concerns with the improved design's ability to meet the requirements on the list of customer needs. This exercise was valuable because it provided a means for discussing areas of disagreement nonemotionally and also indicated where remaining design improve-

Comparison Criteria	Reference Design	Competitive Design A	Competitive Design B	* *	* *	First Raychem Design	Improved Raychem Design
*		S	–	S	–	S	S
*		–	S	–	S	S	S
Technical Requirement xxx		S	–	S	–	S	+
Technical Requirement xxx		–	–	–	–	S	S
Customer Need xxx		+	–	+	–	+	+
Customer Need xxx		S	S	S	S	S	S
Sale Price		–	S	–	S	–	S
*		S	S	S	S	+	+
*		S	–	+	–	–	+
*		S	+	S	+	+	+

Key: + means "clearly better than the reference design"
– means "clearly worse than the reference design"
S means "about the same as the reference design"

FIGURE 2-8. Product comparison table.

ments could be made. The structure that QFD provided to both the project and the interactions among the team members had created an environment where such differences of opinion could be discussed and resolved.

Looking back on the project, the team members advised future development teams to spend more time exploring feasibility and alternative designs based upon in-depth understanding of the customer requirements.

BEYOND THE HOUSE OF QUALITY

After finishing the product planning matrix work, the QFD process was not formally continued. Some QFD charts were completed, however.

Parts and Design Analysis

After finalizing the design, those part features essential to performance (such as dimensions and materials) were identified and prioritized in a smaller matrices. An example appears in Figure 2-9. This matrix, which related technical requirements to part characteristics, is a standard matrix used in QFD work. It is known as a parts planning chart—a Phase Two matrix in the Four-phase system [2], or a C-3 matrix in the Matrix of Matrices system [7].

As a result of the QFD analysis of part features, testing was increased on some of the highest-scoring part characteristics. Designed experiments conducted on part dimensions and material properties increased the team's

				MANDREL				RING		
Relationships Strong = 9 pts Medium = 3 pts Weak = 1 pt			TARGET VALUES →	.xxx in. +/- .xxx	.xxx in. +/- .xxx	xx Rock-well		.xxx in. +/- .xxx	Compound xx-xxx	
			PART FEATURES →	Leaf Wall Thickness	Port End I.D.	Plating Hardness		Ring Wall Thickness	Ring Material	
KEY TECHNICAL REQ'TS	TARGET VALUES		Tech. Req't Importance							
•			•					□		□
•			•		■					
•			•	■	■			■	□	■
Force to Install	x lbs		•		■	■	■			
Force to Remove	x lbs		•	■	■	■	■		■	■
RF Shielding	xxx dB		•		■	■	■			■
•			•	■				□	■	
•			•			□	■			
•			•				■	■		■
Weighted Importance of Part Features				••	171	63	132	••	153	39	••

FIGURE 2-9. Technical requirements vs. part characteristics.

understanding of the range of values required for proper performance. The key part features identified at this stage also formed the basis for quality control procedures established later in development.

To complete the work on the key part features, a subgroup of development and manufacturing members constructed two separate part feature matrices: one for the connector-to-port interface and one for the connector-to-cable interface. Only three technical requirements were common to both matrices, which were 8 × 14 and 10 × 13 items in size. As before, overhead transparencies of blank matrices were used to record the key part features agreed upon by the group and also to record team consensus on relationship weightings.

New importance weightings were established using group discussion for the selected technical requirements. A 1–5 scale was used, with 5 as most important. The group did not simply transfer the calculated technical requirement weightings from the first matrix, as is often recommended. The priorities were re-weighted before proceeding because some very important technical requirements had received relatively low scores. It was not clear why, but some of the technical requirements essential to product success had few important relationships to the items on the customer needs list.

The Raychem Design Review process requires analysis and documentation of product performance limitations and potential product failure modes prior to release of all new products. Accordingly, development engineers completed a design failure mode and effect analysis (FMEA). As background work for the FMEA, the team completed a fault tree analysis [10] for the connector. Fault tree analysis appears in the Four-phase system [2] as part of Phase Two work and is known as the F-2 matrix in the Matrix of Matrices approach [7].

Developing the Manufacturing Process

The component parts of the CATV indoor connector are manufactured by vendors and later assembled by the Telecom production group. Due to intense project timing and resource constraints, the decision was made to develop and document the CATV connector manufacturing process using existing company practices, rather than to continue to experiment with using formal QFD methods. A process flow chart was created, along with detailed part drawings and purchase specifications.

During process development for the connector shell, engineers worked with polymer suppliers at vendor plants to select materials and develop appropriate processing. Purchasing representatives joined the team temporarily at this time for work on setting up visits between development engineers and die casting vendors to discuss production, design, and process alternatives for the metal mandrel part.

A quality plan was drawn up in the form of tables outlining Raychem receiving inspection procedures, documentation and labeling requirements, functional tests, measured characteristics, sampling methods, measurement tools, and relevant document numbers.

In retrospect, the team felt that several initial manufacturing floor problems could have been avoided through formally continuing the use of QFD in process development. The team recommended that future teams use QFD to develop manufacturing processes, rather than stopping after the requirements definition and qualification testing stages. The QFD work on technical requirements and key part features did ease process development work, however, because it resulted in a better understanding of the technical priorities and part tolerances required.

Other Raychem teams have taken QFD further into the development process. QFD has been utilized successfully in manufacturing process design at Raychem in conjunction with designed experiments, statistical quality control (SQC), and statistical process control (SPC).

THE IMPACT OF USING QFD

Timing

Many factors outside of QFD influence how long a development project will take to complete. These include the complexity of the market and customer base as well as of the project and product, the need for materials development, the extent of the existing knowledge base, and the priority and resources dedicated to the project.

In the team's opinion, using QFD improved the product without delaying the schedule. Although several team members had feared that using QFD would lead to major schedule delays, the added time spent gathering and analyzing the customer and technical requirements was regained during the product testing phase due to an improved understanding of the technical priorities.

For the CATV connector team, completing the QFD charts was unfamiliar work that took a long time. This work proceeded most efficiently, however, when significant time blocks, such as half-day meetings, were devoted to completing specific QFD-related tasks. Keeping the charts small was also important in making progress with QFD visible to the team.

The CATV indoor connector, which involved significant materials development work, was formally introduced approximately 2 years after initial inception and 10 months after the start of the QFD study. The first 3 months after starting work with QFD had been spent interviewing customers, completing the elements of the product planning matrix [12], improving the design, and setting up qualification testing.

Product Introduction

The indoor connector project is considered to be an important successful first application of QFD methods within the Telecom Division. Fewer design changes and higher manufacturing yields were experienced at product introduction, since many issues had been clarified during the QFD study.

The indoor connector was extremely well received by CATV customers. It was seen as a real breakthrough in terms of specifically addressing the needs of the indoor application environment with a high-performance, cost-effective product. The connector, while less than perfect at product introduction, was significantly better than it would have been if QFD had not been used. Subsequent improvements have enabled the indoor connector to achieve rapidly growing market acceptance.

Magic?

It is easy to regard QFD as a magic solution to the many problems involved with new product development efforts. In fact, QFD is a methodology that combines elements from many common practices, including customer visits, market research, concurrent product/process engineering, and value engineering. If there is any magic to QFD, it may be in the way it promotes teamwork. QFD provides a structure for involving all disciplines throughout the development process and requires ongoing communication, documentation, and prioritizing to take place.

For the CATV connector team, the QFD process served an important team-building function [5]. QFD clearly defined the topics to be addressed and required consensus on the reasoning behind decisions. As a result, the discussions during team meetings focused on facts and data rather than on opinion. Asking fundamental questions was encouraged, and all team members' viewpoints were considered.

The QFD charts improved project documentation and forced the team members to make data-driven decisions in prioritizing project issues. The discipline of using the charts also encouraged team members to document and distribute copies of their individual work. Matrix updates were published after meetings, along with minutes describing decisions, progress, and action items.

Using QFD cannot ensure that all potential problems will surface at the beginning of a project. Late in the CATV connector development work, a new port size was discovered; a design adjustment was made for the new port in a follow-on product improvement. To accommodate the hundreds of on-off cycles imposed on connectors when customers used them for training or demonstration purposes, a product update was developed and introduced. These possible uses were not foreseen by the development team.

The vision achieved by the CATV connector team through setting "world class" performance values based upon future customer requirements has set goals that will be pursued via incremental product improvements in coming years. The QFD effort is ongoing to understand and respond to new discoveries and to keep listening to the customer voice as it changes with time.

Evaluation by the Project Team

A meeting of the CATV connector team was held to sum up what went right and what went wrong and to gather suggestions for future work. Figures 2-10, 2-11, and 2-12 summarize the three lists brainstormed by the team. High on the list of benefits were improved communication and greater ease in making decisions based upon facts rather than opinions. Key problems were including too much information in the first matrix and starting the QFD process after the design concept had been chosen.

Evaluations by other Raychem teams of their experiences using QFD emphasize in particular the advantages of improved communication and teamwork, and also note the benefits of using QFD to understand the reasons behind customer specifications.

A significant change in resource allocation accompanies the use of QFD. The commitment of a full cross-functional team to frequent long meetings

What Went Right

The Team Process

- QFD forced prioritizing the requirements.
- The process kept group members involved and connected.
- The structure made observations rational rather than emotional.
- QFD helped identify differences of opinion and reach consensus.

The Product

- Matrices displayed the priorities more clearly than lists.
- QFD forced direct technical group contact with customers, early in the project.
- QFD helped identify areas where more customer information was needed.
- QFD forced looking at the competition, point-by-point.
- The product was improved.

FIGURE 2-10. Advantages/insights gained using QFD.

What Went Wrong

The Environment

- Development events and timing squashed the process.
- The design concept had been chosen; relatively few changes could be made.
- The team lacked the buy-in of all members to using QFD.
- Lack of meeting attendance by some members resulted from the lack of buy-in.
- Normal staffing practices did not allow enough resources.

The Tool

- The first matrix was too large.
- We mixed in tests with the technical requirements.
- We did too much numerology on the customer data.
- The process took a long time.
- We were late in distinguishing "world class" performance from required product introduction target values.

FIGURE 2-11. Key problems using QFD.

Suggestions

Customer Interaction

- Apply QFD early, not after choosing the design concept.
- Involve all group members in customer visits and gathering customer data.
- Gather customer data scientifically to get consistency; phrase questions consistently and cover the same topics with each customer.
- Do not announce the product introduction date too early.
- Don't show the product prematurely and encourage competition.

The Tool

- Do a preliminary House of Quality early and then refine it later in the project.
- Use fewer, broader customer needs and technical requirements.

Project Management

- Spend more time on feasibility and alternative designs.
- Clearly identify the target values required to go to market.
- Spend more time understanding interactions among technical requirements.
- Don't expect a shorter time to product introduction, but do expect a better product.

FIGURE 2-12. Suggestions for future work with QFD.

and between-meeting assignments creates heavy resource loading early in the life of a development project. For the CATV connector team, normal staffing practices did not provide the concentrated time needed for conducting the QFD study as well as for accomplishing other scheduled work. As a result, team members devoted many additional unscheduled hours to the project, and conflicts with existing commitments developed. Work on the QFD charts was completed relatively slowly, due to the competition for resources. Despite these difficulties, however, the extra hours spent early in the indoor connector project reduced the need for even higher levels of engineering support later in development work.

INTEGRATING QFD INTO PRODUCT DEVELOPMENT PRACTICES

The benefits achieved in the indoor connector project through the use of QFD, even considering the rather difficult project context and constraints, convinced Telecom technical management to standardize the use of QFD methods within the division's product development practices. The use of QFD is now required in all significant new product development projects within the Telecom Division. In fact, virtually all Telecom development projects, with the exception of the most minor ones, use QFD methods.

The Telecom product development process has changed significantly as a result of the simultaneous introduction of cross-functional development teams [11] and QFD. The cross-functional "core team" concept as it is applied in Telecom requires each development team to have assigned members representing manufacturing, development, and marketing/sales. QFD is employed as the vehicle through which the core team members communicate, gather information, and link their efforts together.

In this context, all functional groups involved in product development have direct access to customers. This approach has replaced the former Telecom product development process, where a small group of development engineers worked in isolation on instructions received from sales or marketing and then handed the design over to manufacturing as a final step.

Building upon the experience gained during the CATV connector project, Telecom development teams have refined their approaches to gathering information from customers and to managing the information that is entered into matrices. The result has been better customer data and smaller, more focused matrices. Using QFD has become easier with practice.

Other Raychem groups have also incorporated QFD elements into project review milestones. In some cases, the use of QFD has quickly revealed that key customer requirements did not correspond to internal objectives or capabilities. These projects were terminated, saving the company much time and expense.

At Raychem, QFD is viewd by senior management as an essential tool for success with new product development. QFD elements are incorporated in the company's New Product Introduction Guidelines, and the use of QFD charts and tables in the Raychem Design Review process is part of the company's QFD training.

QFD Success Factors

For maximum effectiveness, QFD requires support from many other tools, skills, processes, and systems [8]. Company culture also plays an important role, because it must supply an underlying belief system whose values reinforce the practices needed for success with QFD.

Throughout the CATV connector development project, the team utilized a wide range of tools associated with TQM and concurrent engineering. Affinity and tree diagrams assisted in organizing information; benchmarking provided information on how well existing products fulfilled customer needs and the related technical requirements; strategy of experimentation yielded efficient experimental designs for determining target values; fault tree analysis and FMEA provided information on potential product weaknesses; principles of design for manufacture and assembly (DFM/DFA) were applied in work with vendors.

Other teams using QFD in product development work within Raychem have gained important benefits by employing cause-and-effect diagrams, project management techniques, focus groups, conjoint analysis, computer product and process modeling, statistical process control (SPC), and statistical quality control (SQC). Of these supporting tools, designed experiments have played a particularly important role in verifying the strength of product/process interactions, determining product and process target values, evaluating competitive products, and quantitatively comparing potential design solutions.

The individual skills of team members and project team leaders are another important ingredient for success with using QFD. Relevant skills include product development experience, technical expertise, problem solving, teamwork, communication, interpersonal skills, meeting management, listening skills, and project management ability, as well as QFD chart knowledge.

Systems and processes also play a major role in supporting QFD. The Raychem Design Review process and the New Product Introduction Guidelines form the framework within which QFD is used by development teams. These processes outline a phased approach to product development work and define the required review, approval, and documentation of major product development decision points.

In areas of Raychem that have management systems structured around cross-functional teams, the use of QFD has proceeded more smoothly than in areas of the company with comparatively little team experience. The active presence of a team-based quality improvement system, for example, generates a high level of experience in working with multifunctional teams, establishes a problem-solving environment, and fosters growth of the customer-focused philosophy that must underlie the integration of QFD into company business practices.

In addition to support from systems and processes, achieving lasting success with QFD requires a TQM-oriented company culture that values teamwork, customer focus, and the use of structured methods.

Work is ongoing, with leadership from Raychem senior management, to establish training [8], to fully assimilate QFD elements into the product development process company-wide, and to instill a strongly customer-focused philosophy at all levels within the company.

QFD as a Change Agent

Within the Telecom Division, management standardization of product development to incorporate QFD methods has placed QFD in the role of a catalyst for change. The drive to use QFD methods in gathering customer voice data has led to the institution of programmatic customer visits [9]. Today, visits to customers by team members representing all functional departments are common; a product development project leader would not think of embarking upon a new project without gathering customer data.

Internal partnerships among functional departments are strengthening as a result of membership on cross-functional development teams. The use of other TQM tools required for success with QFD is increasing, and employee teamwork skills are growing. Telecom Division senior managers have developed and deployed a TQM program to train employees in a uniform approach to improving customer satisfaction.

CHAPTER SUMMARY

The CATV connector team members as well as management strongly believe that customer satisfaction was significantly higher as a result of using QFD. The process of using a multifunctional team to obtain and analyze the customer voice resulted in improved communication and a better product. Although QFD was applied well after the start of development work, many benefits were derived. The team recommends strongly, however, that QFD be used from the very beginning in future projects.

As a result of the first success with using QFD in the CATV connector project, Telecom Division technical management has standardized the division's new product development process to require a QFD approach on all major new projects. Work is in progress to standardize product development practices company-wide, to include QFD elements.

For maximum effectiveness, QFD requires a support structure formed by tools, project team member skills, and company processes and systems. A TQM-oriented company culture and philosophy form the foundation for sustained success with QFD.

References
1. Akao, Yoji, ed. 1990. *Quality Function Deployment: Integrating Customer Requirements into Product Design*. Norwalk, Conn.: Productivity Press.
2. American Supplier Institute. 1987. *Quality Function Deployment*. Dearborn, Mich.: ASI Press.
3. Daniels, William R. 1986. *A Manager's Guide to Using Task-Force Meetings*. San Diego, Calif.: Pfeiffer & Company.
4. Brassard, Michael. 1989. *The Memory Jogger Plus: Featuring the Seven Management and Planning Tools*. Methuen, Mass.: GOAL/QPC.
5. Graham, Robert J. 1989. *Project Management As If People Mattered*. Bala Cynwyd, Pa.: Primavera Press.
6. Hauser, John R, and Don Clausing. 1988. The house of quality. *Harvard Business Review*, May–June 1988, pp. 63–73.
7. King, Bob. 1987. *Better Designs in Half the Time, First Ed.* Methuen, Mass.: GOAL/QPC. 1987.
8. Liner, Marilyn. 1992. Developing company-specific QFD training: a customer-driven approach. *Transactions from 'The Fourth Symposium on Quality Function Deployment.'* Methuen, Mass.: GOAL/QPC.
9. McQuarrie, Edward F., and Shelby H. McIntyre. 1992. *The Customer Visit: An Emerging Practice in Business-to-Business Marketing*. Cambridge, Mass.: Marketing Science Institute.
10. Mizuno, Shigeru, ed. 1988. *Management for Quality Improvement: The 7 New QC Tools*. Norwalk, Conn.: Productivity Press.
11. Rosenau, Milton D., Jr. 1990. *Faster New Product Development: Getting the Right Product to Market Quickly*. New York, N.Y.: American Management Association.
12. Technicomp Inc. 1989. *Quality Function Deployment Application Guide*. Cleveland, Ohio.

3

Concurrent Engineering Delivers on its Promises: Hewlett Packard's 34401A Multimeter

Robert A. Williams
Hewlett Packard

INTRODUCTION

Advocates of concurrent engineering have for years promoted the concept of developing a product with a cross-functional team of engineers and product support areas. The general idea is to have these individuals involved simultaneously at the front end of a product's life cycle. The promises of the concurrent engineering concept have been touted extensively in many publications since the early 1980s and include such high-level advantages as:

1. Faster time to market.
2. Reduced manufacturing costs.
3. Higher quality levels.

In short, concurrent engineering promises to deliver an increased competitive advantage to the company that not only hears its message, but sincerely applies the principles contained within its scope. So the question is whether a product development team avidly pursuing concurrent engineering can truly expect these results. Also, will this concept of the 1980s and 1990s really provide the kind of breakthrough that companies need to compete in the years ahead? One way to find out is to examine a case history of a product development team that avidly followed the tenets of the concurrent engineering concept. The development of the Hewlett Packard 34401A multimeter demonstrates the progress that can be realized when the necessary ingredients are in place and the people involved are given a chance to excel. The HP

34401A represents a significant advance in price, performance, and value in the test and measurement market. This chapter presents the method by which one HP team integrated tools such as design for manufacture and assembly (DFMA), quality function deployment (QFD), and activity-based costing (ABC) into the simultaneous engineering process, to give the customer what he or she really wants at a competitive price.

MARKET BACKGROUND

First, it is important to briefly characterize the nature of HP Loveland's product offerings and the complexion of its markets. HP Loveland designs, manufactures, and markets a variety of electronic products for use in the test and measurement market. This $5 billion market is presently characterized by slow growth and increased competition from both Europe and the Pacific Rim. Parts of the instrument market are tied to defense-related activities, which have softened since the demise of post–World War II communism. The digital multimeter segment of the test and measurement market comprises about $300 million. Products in this mature market range in price from a few hundred dollars to several thousand dollars, based on performance and capability. The 34401A development challenge was to deliver the performance of $3,000 to $5,000 instruments at a $1,000 price. Meeting this objective was a vital part of our overall business strategy. To do this and still provide growth through satisfactory profit would require a comprehensive and fresh approach to the product development cycle.

HP34401A TEAM FORMATION

The formation of the 34401A team was done in a number of phases rather than as an immediate and deliberate attempt to define a huge, simultaneous engineering team. The initial staffing consisted of a project manager, a mechanical R&D engineer, two electrical R&D engineers, and a manufacturing engineer to review alternatives to the mechanical approach to the product. Four more R&D engineers and a marketing engineer were added shortly to complete phase 1. The second phase involved the assignment of several manufacturing engineers and, subsequently, a manufacturing project manager. Phase 3 followed quickly when two events occurred. The first was the collocation of the entire development team, and the second was when management expanded the boundaries so that the team could function as a "virtual Greenfield" effort. The team, consisting of R&D, Manufacturing, and Marketing was now in the same location, enabling easier communication and alignment of purpose. Studies have proven that, if even small distances exist between team members, communication is compromised. By collocating the team, these types of barriers to communication were removed.

As a part of the second event, the team was allowed by upper management to challenge the existing financial systems and processes that were in place. The R&D and Manufacturing project managers worked together to develop a strategy concerning financial and accounting-related issues. Traditional cost accounting would bear tough scrutiny on the 34401A project.

LEARNING TO WORK TOGETHER

Phases 1 and 2 had incorporated some design for assembly and some design for manufacture focus. There was also extensive activity in the area of market research. But not until phase 3, when the entire team was geographically together, did we realize that there was something lacking in our product development culture. That something was, quite simply, the skill of working together as a cross-functional team. The R&D portion of the team had worked together before on similarly challenging projects. However, most of the manufacturing staff had not previously worked together and also had to quickly catch up with the team momentum. They were now thrust into a new daily interaction with each other, and we quickly discovered that working together as a cross-functional team is a progressively learned skill. Unfortunately, the technical community is typically not trained in academia to function as team members. Engineers are especially vulnerable to individual competition in undergraduate- and even graduate-level education. As a result, they can sometimes bring a maverick individualism into the workplace that disrupts common sense endeavor, such as concurrent engineering. To overcome some of these challenges, the project team took some collective training in such areas as team development, conflict management, and personality type analysis. We needed to discover how to constructively handle the inevitable conflicts that arise in product development. Also, team development models and the Myers-Briggs personality type analysis were utilized. These analyses helped us learn how to benefit from diversity instead of working against it.

THE TOOLS

Aside from extensive market research, the HP 34401A project primarily used three key tools in the course of its development. QFD, ABC, and DFMA were the primary team tools used to focus on the areas we felt were most important to the success of the product.

Quality Function Deployment (QFD)

As mentioned before, extensive market research had been conducted prior to team collocation. This effort continued with phone surveys, key customer visits, focus groups, mail surveys, and interviews with internal users of multimeters. The information gleaned from this research was compiled by the team and was analyzed using QFD. QFD was valuable to the team in a number of ways. First, it was a tool that allowed us to focus on real customer needs. The all important Voice of the Customer was a concept that helped us navigate through times of disagreement. Second, QFD allowed the requirements asked for by the Voice of the Customer to be translated into measurable manufacturing requirements. For example, one of the clear considerations of future customers was the availability of the instrument. A specific objective of 24-hour turnaround, from order receipt to the shipment of the multimeter, drove manufacturing to develop new processes that could deliver at that throughput level. Third, QFD became a common thread allowing Marketing, Manufacturing, and R&D functions to actually optimize the overall business objectives. It provided the necessary alignment of goals and objectives for the whole product team. The benefits of using the QFD Process can be summed up by the following lists.

It was used to:

1. Focus on delivering "just enough" functionality to meet customer needs.
2. Identify customer "excitement features."
3. Document decisions, considerations, and rationale.
4. Make trade-offs between alternate scenarios.
5. Communicate expectations and rationale.
6. Disseminate information between team members.
7. Do front-end decision making.

Its output was a document that:

1. Didn't presume a solution.
2. Captured financial goals and boundary conditions.
3. Captured factory performance metrics in addition to cost.
4. Embodied the product development vision.
5. Provided a common communication medium with others.

Activity Based Costing (ABC)

In the past, products had been developed using a bottom-up approach. That is, they were first designed with rough customer goals in mind, then costed, and then priced. The HP 34401A reversed that approach. With our market research in hand and the QFD done, we knew exactly what the price needed to be. Now it was a matter of costing the product. The design would have to fit these hard cost goals.

Much has been written about the virtues of ABC in recent years. HP's form of ABC had been in place about 4 years at the time of the 34401A project. At that time there were in excess of 15 manufacturing process cost drivers in place. However, our ABC system just wasn't sensitive enough to the changes we proposed. It captured the status quo very well, but it became obvious that it would not represent the true costs of the processes and improvements being considered. Its limiting factor was that it assumed the use of existing processes and allocations. For example, there were still some non-process–related costs that were being allocated across complete product lines. This was being done without regard for such things as the difference in engineering support requirements or differences in field warranty rate. Therefore, we needed an enhanced financial model to characterize this specific product as if it were the only product in a start-up type of business. Such a model was developed by the team responsible for breakthrough costing, including the finance department partners. The objectives of this financial model were to:

1. Lay the foundation for modeling the 34401A profitability.
2. Evaluate different manufacturing options, processes, and volume-sensitive scenarios.
3. Document assumptions.
4. Provide feedback in less than 2 hours.

Basically, we wanted to be able to make quick, accurate decisions based on the product's stand-alone true costs while shedding old traditional cost burdens. Design trade-offs could be quickly evaluated based on the off-line models. For example, we closely examined areas such as material stock and pull. New methods were developed to have suppliers deliver directly to the line in a "weekly" just-in-time (JIT) process. We also improved the reliability of the unit to avoid costs for troubleshooting and technician allocations. This was done by making the design more robust and then verified with exhaustive reliability testing. We also concentrated on selecting higher-quality materials and components.

Design for Manufacture and Assembly

Before outlining DFMA on the 34401A, it is important to understand some of the prior experience that HP Loveland had with design for assembly, design for manufacture, and concurrent engineering. In the mid 1980s, initial DFMA efforts at Loveland centered around redesigns of existing products to enhance their manufacturability. There was a reserved opinion about the credibility of any "new" approach, such as DFMA, to have a favorable impact on new product development. Most managers saw it as a threat to meeting their schedules. Because of these conditions, the redesign route was chosen to "test" the power that DFMA could have. In a previous writing by the author, the experience was summarized with an assuring message. The conclusion was that, even on existing products with far more constraints to design than on a new project, DFMA efforts are not only worth it, but they provide excellent competitive results. Production cost savings on these redesign projects ranged anywhere from a low of 5% to a high of 18% [1].

From that point, HP Loveland embarked on several other DFMA efforts in new product development and achieved significant success in the reduction of parts, assembly time, number of operations, and types of fasteners used in their products (see Figure 3-1). Product development during this time took the form of one to three manufacturing engineers interacting frequently with members of the R&D team. On occasion, one of these manufacturing people was collocated with the R&D team for several months as the product evolved in the lab. The assembly improvements during this time period were primarily due to a heightened awareness of DFMA by the designers. This awareness was accomplished by training in the basics contained in DFMA concepts. (One retrospective lesson learned from these early successes was that the cost-cutting power of good design through DFMA was enough to create excess assemblers on released product lines. We did not adequately attend to the long-range labor planning affected by these improvements. We were

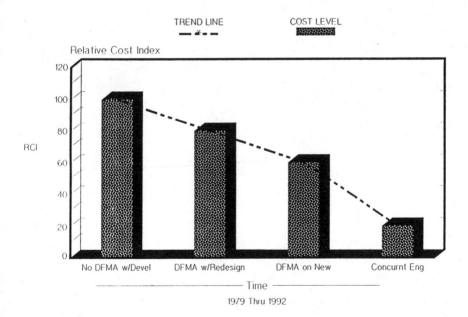

FIGURE 3-1. Relative improvement using DFMA and moving toward concurrent engineering.

50   Successful Implementation of Concurrent Engineering

eventually faced with the resizing or redeployment issues of the people that used to put these products together, and thereby lost some of the cost improvements while doing so.)

On the HP 34401A, every piece of the product was analyzed using the DFMA methodology and software from Boothroyd-Dewhurst, Inc. As a result, there are many DFMA-related features on the 34401A, but probably the most innovative of those is the input connection scheme. Previous multimeters from both HP and its competitors utilized point-to-point wiring, from the terminal receptors to the printed circuit board. Because analog measurement requires high impedance, Teflon wire was hand soldered between these two points. That meant a total of ten wires, each with two hand soldering operations at final assembly. A clever design, consisting of formed copper tubing of the required diameter for a "banana jack," coupled with a high-temperature plastic housing, allowed these terminations to be made at the wave solder machine rather than having to hand solder at final assembly (see Figure 3-2). Also of particular note is the fact that the entire front panel assembles with no screws (see Figures 3-3 through 3-18 for details of these types of features; comparative results are shown in Figure 3-1).

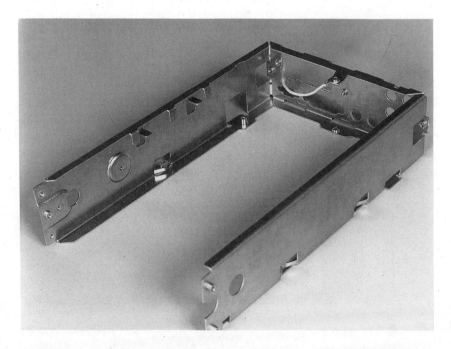

FIGURE 3-2. The base chassis; wraparound plated steel design. Note built-in features for transformer wire routing and built-in hooks to accept slots in the printed circuit board.

FIGURE 3-3. Analog-printed circuit board showing mixed technology and an extra terminal block with back portion of plastic disassembled to show design of block.

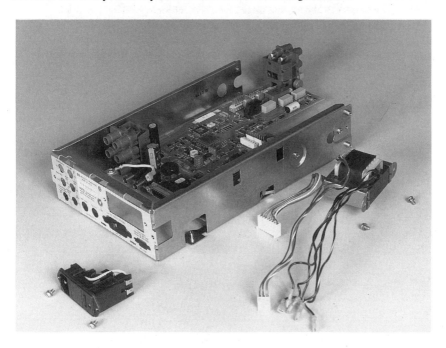

FIGURE 3-4. Interior assembly coming together. PC board partially slid into chassis. Fuse, power switch, and AC input presented as one part at final assembly station.

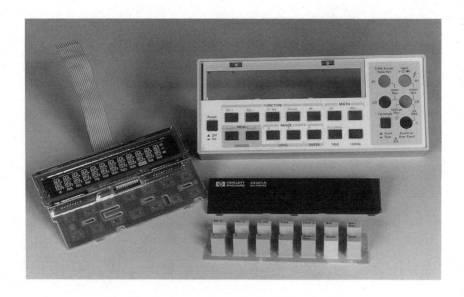

FIGURE 3-5. Components of the front panel assembly. Vacuum fluorescent display presoldered onto display PC board, one-piece conductive rubber keypad, snap-in display window, and front panel bezel.

FIGURE 3-6. View of the backside of completed front panel subassembly. Note built-in cantilevered PC board material that snaps z-axis assembly into place and secures.

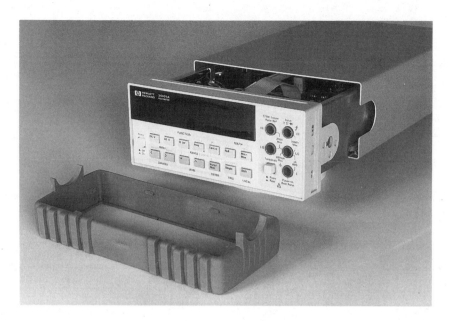

FIGURE 3-7. Internal subassembly sliding into vinyl clad aluminum cover. Rubber bezel stretches easily over plastic front bezel.

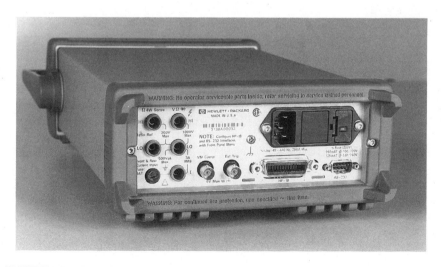

FIGURE 3-8. Rear view of completed 34401A instrument.

FIGURE 3-9. View of assembly area (on left) and test area (on right).

FIGURE 3-10. 34401A in process of assembly. Note close proximity of the parts and total inventory.

FIGURE 3-11. Close up of 34401A multimeter shield being attached with just one screw.

FIGURE 3-12. Multimeter ready to pass to the right for final parts to be added, such as outer sleeve, rear bezel, and rubber bezels.

FIGURE 3-13. Close up showing partially assembled multimeter, limited number of tools, one screw type, and close proximity of parts in their gravity racks.

Many of the ideas for these features and subassemblies were conceived well before the bulk of the team had arrived. But it should be emphasized here that it took a cross-functional team effort to identify producible designs, materials, and the correct suppliers to make the ideas work. The key deliverable of any DFMA effort is a significantly reduced part count. A lower part count allowed us the freedom to try some new manufacturing processes.

A TOUR OF THE FINAL ASSEMBLY AREA

One of the most efficient ways to discover the relative success of a concurrent engineering effort is to take a look at the product's final assembly area. A visit to the 34401A's final assembly cell is visual proof of long-term thinking, thorough planning, and reliable design. To begin, we see a multimeter that can be assembled manually by one person in just over 6 minutes, compared with 20 minutes for the unit that it replaces. The product is then placed in a test system directly behind the main assembly area. After the unit is temper-

FIGURE 3-14. The small and efficient 34401A assembly cell from approximately 30 feet away, showing view as the stocking supplier would see it on approach. Parts are placed in the racks slanted toward assembly side, and empty containers from last week's build are taken back to the supplier.

ature stabilized, zero calibration is accomplished, followed by full-scale calibration and performance verification.

We also observe that a week's worth of parts resides in a space of less than 300 square feet and that the production line is designed so that most part suppliers can bring parts directly onto the production floor, stock them, and remove empty totes and boxes from last week's build. Industrial dollies are rolled from the suppliers' trucks right into the final assembly cell. The environmental problems and time-related costs of dealing with corrugated boxes, packing material, and the whole issue of detrashing has virtually been eliminated by these recyclable totes and durable dollies. Visitors also see production and assembly personnel ordering next week's required material from suppliers via "fax kanban" through a fax machine installed in production, without the need for a material buyer on a regular basis. Open purchase orders with key suppliers have been negotiated long ago by a buyer. The production personnel give regular quality feedback to the suppliers via phone

FIGURE 3-15. A closer view of the inventory racks, showing a typical dolly used for the larger parts.

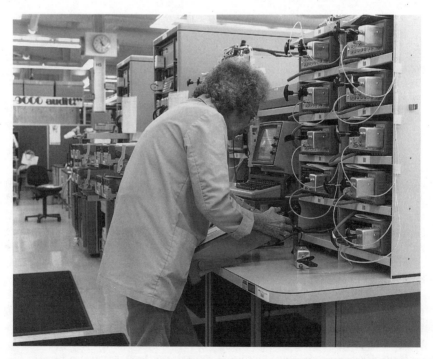

FIGURE 3-16. Preparing a newly assembled unit for temperature stabilization, calibration, and testing.

FIGURE 3-17. Oblique view of test stations for the 34401A.

or once a week when deliveries are made directly to the line. With only 18 parts to keep track of, demand variations can be responded to easily, by increasing or decreasing next week's orders. Inventory levels are immediately visible and are monitored once per week and then compared to orders to calculate future required part quantities. Inventory turns are greatly enhanced, and material has a much higher throughput than previous designs. Typical instruments, for example, have approximately 60 to 75 days worth of inventory at any given time. The 34401A averages 2 to 4 days worth of inventory. Production engineers have limited involvement, since turn-on rates are so high and field warranty rates are lower than previous instruments.

The key point is that part count drives virtually all of the downstream processes in manufacturing. Without development tools, particularly DFMA, these competitive advantages could not be realized. Without the concurrent product development team working together early on, cost benefits such as these could not have been captured.

FIGURE 3-18. The "Fax Kanban" being used to send the week's orders. Regular material buyer is not needed because the line personnel do the releases of material with the supplier partner.

DEALING WITH NEW TECHNOLOGY AND ADDITIONAL CONCURRENT ENGINEERING

Additional cross-functional efforts tackled by the team involved characterization of new printed circuit board assembly techniques, concurrent custom-integrated circuit part and process development, and design and process characterization for process capability.

First, the technical product definition that evolved in phase 1 and then later as QFD was rigorously applied made one thing obvious. We would be forced into a serious consideration of using at least some degree of surface mount technology (SMT). To the casual observer this probably would not be so impressive, given the broad acceptance of SMT. However, one must consider that a system multimeter capable of measuring one millionth of a volt requires exceptional cleanliness in the PCB and PCA processes. We had never done a precision analog measurement board using mixed through-hole and surface mount technology before now. The evaluation of variables such as solder mask type, platings, trace widths, laminate type, solder paste cleanliness, post-reflow cleaning techniques, and other mechanical part innovations, made for a project within a project. A manufacturing process engineer on the team was assigned the task of investigating many technical alternatives and then pulling together the peripheral resources necessary for implementation. At the same time this person interfaced with the surface mount factory team particularly to convey reliability goals of the product.

The second area involving technology was the challenge of developing a custom integrated circuit (IC) for which there was no existing IC process. We had to work concurrently with our in house IC facility and together, develop not only the device, but the process for manufacturing it. This was obviously a crucial leg in the product development cycle because it involved a great deal of the specifications required for higher performance. It was also vital in the quest to reduce the number of parts on the printed circuit board, which in turn helped reduce costs. The IC not only had to be developed, but had to meet aggressive producibility goals that were spelled out in the QFD effort. Process capabilities (Cp k) were set at 1.33 minimum (see Figure 3-19 for graphic of process capability).

The third area involving a new approach was the effort to ensure that all processes used and their parts could meet the 1.33 minimum process capability number. This kept members of the manufacturing portion of the team very busy. The same design care and exhaustive testing philosophy had to pervade every part and every supplier who made the part. The producibility of the total product is only as good as the multiplied producibility of each of its individual components.

$$C_p = \frac{\text{SPECIFICATION WIDTH}}{\text{PROCESS WIDTH (6s)}}$$

$$C_{pk} = \text{MIN. }(C_{pu}, C_{pl})$$

$$C_{pu} = \frac{USL - \bar{\bar{X}}}{3s}$$

$$C_{pl} = \frac{\bar{\bar{X}} - LSL}{3s}$$

(a)

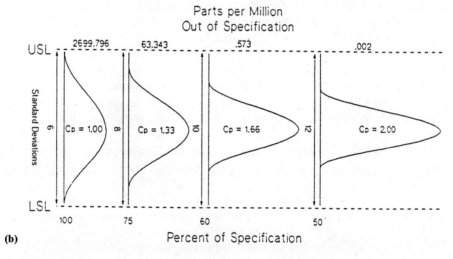

(b)

FIGURE 3-19. (a) Process capability. (b) Process capabilty indices.

RESULTS DELIVERED BY CONCURRENT ENGINEERING

At this point the question to be asked is, "Did concurrent engineering deliver its promises to the HP 34401A project?" The following table contains an abridged list of the promises of concurrent engineering, along with the comparative data from previous HP products.

CE Metric	HP 34401A (%)	Previous Generation (%)	Previous Similar Product (%)
Material $	80	100	200
Nonmaterial $	55	100	250
Assembly time	37	100	210
Average repair time	33	100	400
No. of mechanical parts	30	100	190
No. of fasteners	31	100	172
No. of fastener types	8	100	85
No. of connects, disconnects, adjustments	36	100	120
Final assembly part count	40	100	153
Total parts	68	100	190
Total part #'s	77	100	150
No. of suppliers	70	100	N/A
Inventory days	4	100	100
Throughput	100	100	100
1st year engineering changes	0	100	58

SUMMARY

One of the early keys to the success of the 34401A was the high level of support received from the entire management chain. They gave us the freedom and ability to analyze the total system in order to find where true costs were originating, and they also gave us the necessary infrastructure to form a cross-functional collocated team. There was an overall project manager assigned up front. We were given a basic product development charter, informed of general boundary conditions, and then afforded the freedom to execute.

Also, the success is due in large part to three additional elements:

1. Extensive use of market research.
2. Top-down approach (start with cost and work down).
3. Cross-functional, collocated team effort.

We found a great deal of satisfaction in knowing that we had not taken the easy road to product definition and development. The satisfaction comes when we see the product eagerly received in the marketplace, selling well ahead of projected marketing ramps. Satisfaction also comes when you see the team on the production floor enjoying what they do. They enjoy what they are doing because they have contributed valuable inputs to the project in the early stages. They aren't plagued by problems caused by lack of attention to details. Their involvement added the benefits of rapid response to changing order demands. We see the engineering support function able to devote their time to other value-added projects. This is possible because they aren't fighting avoidable problems on the line or writing engineering change orders. At the time of this writing (15 months after production release), we have not had a single design-related engineering change order written in order to keep the line shipping the product.

Finally, we found that concurrent engineering, if genuinely done, is just plain hard work. As mentioned before, working together as a cross-functional team is a skill that eludes many U.S. firms. Our individualism had to be throttled to the extent where we could be a cooperative contributor to our team and yet still retain our individual creativity. Any way you analyze it, to get highly trained, intelligent, and competitive people to work together on a daily basis is a skill that most companies wish they could just order by way of a decree. But it can be realized only when there is firm resolve, coupled with repeated trial and failure. Only then will the skill be developed. The intrinsic benefits of that developed skill are leverageable into the future, and, if implemented properly, the concurrent engineering effort will again deliver on its promises.

ACKNOWLEDGMENTS

The author gratefully acknowledges the counsel of Ed Leon and Scott Stever during the organization, compilation, and initial editing phases of this chapter. Their insight, encouragement, and time invested was vital in the preparation of this material.

Thanks and congratulations also go to everyone who had a part in the 34401A project. Their patience and determination to test new ideas allowed HP to achieve a very noteworthy product success in an extremely competitive global market.

References
1. Corbett, Dooner, Meleka, and Pym. 1991. *Design for Manufacture—Strategies, Principles, and Techniques*, pps. 188–193. Addison-Wesley Publishers Ltd.

4
A Team Approach to Concurrent Engineering: A Case Study

Steve Zelenak
Hach Company

Michael A. Ridenour
Hach Company

Dr. Wade O. Troxell
Colorado State University

INTRODUCTION

Team-based efforts to increase the competitiveness of U.S. companies are often a traumatic experience. An individual's role may change dramatically. The boundaries that define prerogatives of the team are not always in place or understood. While the individual roles of each of the members are clear when defined by earlier corporate structure and tools, as the team becomes a separate entity roles are much less clear. Even corporations dedicated fully to the concepts described by buzzwords such as "empowerment" and "self-directed cross-functional teams" can benefit by a clearly understood corporate structure. In a team-based environment, the "hows" of implementing new products—the operational aspects of product development—are the team's responsibilities. Management is free to focus on strategic issues and to provide the resource support necessary for successful teams.

Today's vocal proponents of the team approach have provided a seeming plethora of successful implementation stories. These stories provide convincing arguments for the general application of teams to a corporation.

These same stories often ignore common problems associated with such a transition and do not provide a useful set of analytical tools that could alleviate much of the ensuing trauma. As the corporate structure is in flux, the boundaries between those who have the role of making strategic decisions and those concerned with making operational decisions are often undefined or misunderstood. The ability of upper management to define where different decisions should be made and to communicate those definitions can eleviate much of the floundering that can occur when teams are formed.

This discussion focuses on a case study of team management in product development and a method used for defining the communication and decision-making structures that surround a team. Once a structure is understood, plans to fit teams into the company can be formed with the goal of optimizing resources. While such plans still must be formed based on the individual managers' knowledge and feel for their companies, clear and unbiased information would be invaluable. This study offers the use of a well-known engineering tool—the Petri Net to study specific corporate decision making and communications.

Much care must be taken when integrating the team management approach to project management in a company. Ideally, the teams should involve all functional departments of the company (cross-functional), and they must be given enough decision-making authority to perform the task they have been asked to complete (self-directed). Without the proper authority, teams become nothing more than committees who recommend action but have no means by which to implement those actions. At the same time, teams do require clear direction from management on topics such as resources, goals, objectives, and so on.

In the right environment, teams can make projects move quickly. They can increase productivity, increase quality, and reduce time to market. Yet how to determine what the proper environment is, how to nurture that environment in the company, and how to find its bounds within the company are issues that are very difficult to decide.

U.S. corporate culture and, more basically, society limits the amount of the team-building philosophy that can be directly imported from Japan. Americans are raised to be independent, entrepreneurial, and direct. It is important for a company to communicate to the team where the boundaries lie and what the lines of communication are. The definition of these boundaries addresses the questions that come from misconceptions and miscommunications (e.g., Who's in charge? Who has the right to overturn a team decision? Why does management keep interfering?).

By establishing an environment where communication and decision-making structures are well defined, teams learn what the major limitations of the team in a corporation are. The goal should be to manage "the entrepreneurs" inside a corporation for the benefit of the company.

The example used in this discussion is based on a new product development process within Hach Company, a manufacturer of scientific instrumentation.

TEAM APPROACH TO PRODUCT DEVELOPMENT

Companies have applied the team approach to all aspects of operations recently. The concept has gained considerable momentum because teams have the ability to perform functions with more confidence of success than individuals can. If the team is comprised of the appropriate members, very few outside decision-making resources are needed. One of the best aspects of teams is that the concept is easily applied to product development while complementing concurrent engineering practices.

In the right environment, teams can make projects move quickly. They can increase productivity, increase quality, and reduce time to market. However, companies should be aware that, while projects must be managed, the management of teams requires skills and tools that are not always those used by traditional engineering managers. No team will function effectively if they are not given a reasonable amount of autonomy within the company. Often, the more external forces exerted on a team, the less apt the team will be to make decisions and to act on them. This will undermine the original intent of the team approach and can ultimately cause the product development process to move slowly.

The concept of managing projects versus the management of teams is somewhat of a philosophical topic. It does not require great leaps of logic to understand why managing teams is difficult. It is especially true in research or new product development that many of the people involved, by their natures, are not necessarily "team players." Some of the very traits that make them successful in being innovative and all the more valuable to a project make the manager's role especially difficult. An innovator in any field is often knowledgeable, opinionated, stubborn, confident to the point of being obnoxious, and self-centered and, at times, feels that management is an impediment to the completion of what he or she has determined to be the most important tasks to be accomplished. An analogy relating such human character traits to those of a cat does not seem far fetched. While such an individual is difficult enough to herd towards a goal, herding a group of such individuals (the cats) towards a common goal can seem to be an insurmountable task. Such a person, or cat, unless held down, rebels in a restricted environment. These same individuals, working in a widely bounded environment, are capable of remarkable feats. That bounds exist is not counterproductive in even the most liberal team environment, as long as they are matched to the requirements of the corporation and the individuals involved. Defining those bounds so that they are clear requires an understanding of the true corporate structure and lines of communication. A representative model of any system is a valuable aid in understanding that system. A system in which a large number of communications or processes occur simultaneously is especially difficult to model as a whole. While this case study does show many facets of a case history in product development at Hach Company, it also proposes the use of a promising analytical tool for modeling and understanding corporate boundaries and communications.

PETRI NETS AS A TOOL IN PRODUCT DEVELOPMENT

One of the tools available to be applied to analyzing structures where concurrent communications take place is the Petri Net. As this study shows, there is a clear coupling between the communications flow and the decision-making structure of a corporation. Petri Nets are graphical and mathematical modeling tools similar to flow charts that have traditionally (since the initial description of them in 1962 [1]) been applied to systems such as chemical processes [2], software communication protocols [3], and basic manufacturing systems. Recently, they have been applied with some success to concurrent systems in modern manufacturing. Two examples of this are in the scheduling of job shop systems [4] and the modeling and scheduling of flexible manufacturing systems [5].

This case study describes the use of Petri Nets as a tool to model human interactions within a corporation. Even the most complex of models describing human interfaces are not entirely accurate. Human interactions are too complex and spontaneous to hope to describe and predict with certainty. The method used in this study develops a simple model to be used by a manager or team in understanding some of the basic communication lines and decision-making boundaries that they operate within. The use of such an analytical tool can help to lessen some of the emotional responses occurring when changes occur to the structure. In fact, by understanding the natural boundaries of a group, natural improvements in methods of communication or management can surface.

On the surface, flowcharts, Gantt charts, and Pert charts give the same information as Petri Nets. A Petri Net is more clear graphically than the standard flowchart, especially for showing temporal relationships. The Gantt charts and Pert charts are traditionally understood as being unidirectional,

A Team Approach to Concurrent Engineering   69

linear charting tools. Paths shown using Petri Nets can show operations that might have more than one iteration based on conditional tests. This gives the ability to simply show "loops," which is one of the great strengths of flowcharts. Therefore, a Petri Net merges the strengths of the other three methods into a single analysis tool.

In studies of systems such as the Job Shop Problem, the Petri Net was shown to be adequate by limiting the rules defining it and also by adding temporal relationships to the events being modeled. A temporally based Petri Net was the starting point for this study. As described below, the Simple Timed Petri Net protocol used gave insight into the team communications that led directly to improved lines of communication. Also especially of interest was the fact that the study confirmed the philosophy of concurrent engineering in general by showing a clear comparison to serial product development.

SIMPLIFIED TIMED PETRI NET

The mechanism for the study was a Simplified Timed Petri Net (STPN). A STPN has five simple components and a limited set of rules. The components include events, transitions, time for transitions to occur, a marker, and the associated connections or paths between components. These are shown in Figure 4-1.

An event occurs when it is labeled with a marker; when it occurs it sends a marker down one of the paths away from it. The choice of paths can be conditional "ifs," if not conditional, it is arbitrary. A transition occurs when all paths leading to it have received markers. After the time delay built into the transition, it sends a marker down each of the paths leading away from it simultaneously. This gives the ability to show concurrence.

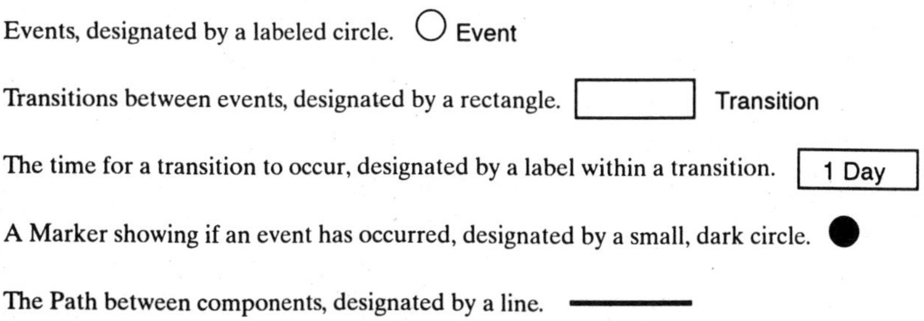

Events, designated by a labeled circle.   ◯ Event

Transitions between events, designated by a rectangle.   ▭ Transition

The time for a transition to occur, designated by a label within a transition. | 1 Day |

A Marker showing if an event has occurred, designated by a small, dark circle.   ●

The Path between components, designated by a line.   ▬

FIGURE 4-1.   The five simple components of a Simplified Timed Petri Net (STPN).

PRODUCT DEVELOPMENT— HACH COMPANY CASE STUDY

Hach Company was founded in 1947 in Ames, Iowa, by Clifford and Kathryn Hach, as a manufacturer of test kits for water hardness. The product line has been expanding since the early years, and in 1992, with sales of $87 million, Hach manufactures an extensive line of laboratory and process equipment as well as test kits and portable laboratories. This case study covers events that occurred in one specific product line of the company. Given the wide range of product types produced at Hach and the many different functions existing at diverse locations, it has been natural that there has been some variance in the way that different areas are managed.

Although intermediate in size, logistically Hach Company offers all the problems associated with a large company: multiple locations, different time zones, different 'first' languages, and different responsibilities to the company organization. Figure 4-2 illustrates the Hach locations and function of each facility.

It is difficult to give a historical perspective on the introduction of new products at Hach Company that provides a reader with a clear understanding of why certain methods were used at a particular point in the company's history. Hach Company started 45 years ago as a husband and wife team working out of garage-like facilities. As the company grew to its present size, employing hundreds of people, management of Research and Development has undergone many changes reflecting the standards of the time.

The steady growth of Hach Company cannot be attributed to an ideal economic climate, as the growth of much of American industry often is. Many of the concepts in place from the earliest days of Hach to drive product development are concepts that are now being "discovered" and "rediscovered" throughout industry. Providing easily used tools and procedures for analytical chemistry that are most often used by non-chemists can only come from understanding and meeting the customer's needs. Innovation often occurred as the result of small groups working together and using whatever methods were necessary to develop products as rapidly as possible that still met the demands of quality dictated by the principles of analytical chemistry. That innovation and quality have been keystones at Hach is shown not only by the variety of products produced by the company, but also the fact that many of the procedures, both novel and routine, that are dictated by the Environmental Protection Agency (EPA) are based on methods and instruments developed at Hach Company.

With growth, age, and a changing environment, product development methods will change and vary from area to area throughout a company. This is also true at Hach Company. At times, product development did seem to be

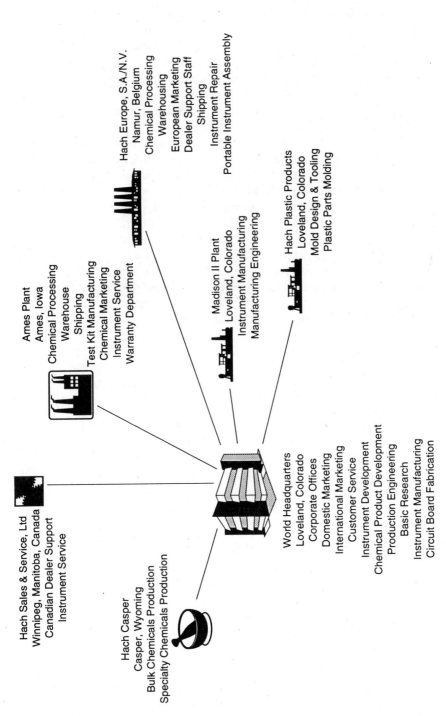

FIGURE 4-2. Hach Company plant locations and function of each facility.

textbook "throw it over the wall" to some. Formalisms were put into place to lessen the impact of such methods on the manufacture of products. While it was not the intent to do so, these formalisms could and at times did change a rapid concurrent development process into a slower, more serial, one. That this could happen was not always caused by the intent of the formalisms, but rather by individual interpretations of those protocols.

New Product Implementation (NPI) protocols were developed at Hach to provide guidelines for the steps necessary to take a product from conception through release. As opposed to the NPI process, which is a formal protocol, some projects historically used little or no formal project planning. This serial design methodology served well for many years and is still practiced in modified form in some parts of the company. In Instrument Operations, however, competitive forces have required the definition of new design methodologies to reduce costs, shorten time to market, and develop families of common products, as opposed to the traditional serial product introduction.

The initial NPI protocol developed at Hach was a rigidly structured, serial development process. It did, however, address the concerns primarily voiced by Manufacturing Engineering about the lack of design for manufacturability of new products. Product designs as presented to Manufacturing often required intensive modification to production tooling before actual production could begin. The intent of this NPI process was to improve the manufacturability of products by requiring documentation of the development process, thus providing a checklist of points common to good manufacturing practice. This protocol also required the NPI Team to administrate the product's manufacture through the first 6 months of production. In these first 6 months of production, Manufacturing personnel were to use TQM techniques to report to the NPI process information.

Simultaneously, the team was attempting to apply steps for continuous improvement, derived in part from those professed by W. E. Deming, to the NPI process. This guided the evolution of the design protocol, resulting in the present system.

The current NPI Plan consists of six phases:

Phase I. Feasibility Model.
Phase II. Engineering Model.
Phase III. Engineering Prototype.
Phase IV. Production Prototype.
Phase V. Production.
Phase VI. Postproduction.

Critical to the success of the current NPI protocol is the ability of the NPI Team to overlap the start of one phase with the end of another or to actually merge two phases into one. The strategy used is dependent upon the complexity of the product being developed as well as the technical ability of the team members. Although it is implied, the NPI protocol does not dictate serial development. It is flexible enough to allow concurrent engineering practices, but this is often a matter of team or manager interpretation. In any case, forces external to the team, such as the availability of resources in other areas or market pressure, at times dictate concurrence. Following is a case study of one NPI Team's approach to new product development, focusing on the necessity of concurrent engineering and product quality.

The team was formed to develop a replacement for an existing product that had been successful for many years but that contained several design flaws. The existing product was bulky, costly, and difficult to manufacture by today's standards. A major goal of the project was to improve the manufacturability of the product, leading to lower unit costs.

The project started when the marketing department requested that the replacement product be developed within 18 months to a point where complementary products development could begin. This meant that the new product had to be ready to enter the production prototype phase at the end of the 18-month window.

A team was established with ten core members and five support members. The core members are:

1. Mechanical engineer (facilitator—headquarters)
2. Quality engineer (headquarters)
3. Purchasing representative (headquarters)
4. Research chemist (headquarters)
5. Process engineer (HPP)
6. Manufacturing line supervisor (MAD II)
7. Product manager (headquarters)
8. Model shop supervisor (headquarters)
9. Manufacturing engineer (MAD II)
10. Service representative (Ames)

The support members are (all located at headquarters):

1. Research engineer
2. Vice-president, Long Range Research
3. Plastics design engineer
4. Electrical engineer
5. Research chemist

The support members are primarily those that were associated with the project during the purely conceptual stage of product development. As part of the NPI protocol, they remain available to the team mainly as informed consultants. They receive all the formal communications from the NPI Team in order to keep themselves somewhat current on the project's status. Because of this, the team receives support from outside of the group on the operational level in two ways. First and most often, the team contacts someone outside of the group with a specific question. The person contacted would be the one with the expertise to assist in solving an immediate problem. The support group participates in this fashion, but also, since they remain informed about the project, are free to offer assistance whenever they feel it is necessary. This provides a continual peer review of the team's work in a very informal way.

There were two brainstorming sessions held in which all members of the project team as well as invited guests developed a preliminary list of product expectations, which provided a good starting point for the core team. One outcome of these brainstorming and definition meetings and especially of Marketing's input was a draft of the formal document describing the product.

Hach Company uses a development memo (DM), which is the living document outlining the attributes, project cost estimations, and other pertinent information to guide the project. This document is periodically updated by the NPI Team as new information is available regarding the project. Drafting this document is the "official" starting point of the project for the core team and took approximately 2 weeks to complete.

The DM is the starting point in life cycle of the NPI Team, and the team is officially disbanded after a postproduction review of the project, product manufacture, and service reports. In many projects, once the team is officially dismembered, it is not unusual for some of the members to spend a fairly large percentage of their time following up on the product. Not only do people such as the manufacturing engineer work with the product in a routine fashion, at times the designers and others are used as consultants in solving problems related to the manufacture of the product.

The phases of the NPI protocol do tend to blend together in a fashion dictated by the project. This alters the level of activity of different members during different phases. Also, the team dynamics do vary as members change. These changes in membership can come from turnover in personnel that leave the company and also from internal rearranging of duties. During a project that goes on for any length of time, such turnover (at least on a small scale) can occur. Some purists of team philosophy say that, when one member of a team leaves and is replaced, the team is altered in a fashion that creates an entirely new team. They would say that the change is immediate and that the early results are negative. This case study shows change that occurred slowly when the team facilitator was replaced. To use a mechanical control analogy, the disturbance was at first slight since the new facilitator attempted to keep the system moving in the direction it had been. Only as the team moved further and further from its goals did the disturbance and response become greater.

The early results of such a change in team membership generally are negative. New members must find how they fit into the team on an informal basis. Informal interaction is the true working process of a team. Until a new member and the team are working smoothly together, the NPI Team is not functioning at its best. Along with this disruption in team dynamics, new membership can bring immediate benefits. New product ideas can be injected into the product, not from a part-time outsider but from a member of the team with a fresh perspective on the project. New members, if they assume an active role, can quickly become acquainted with the operational details of the project. This makes their input especially valuable to the rest of the team, as it is timely and often does not pertain to issues and design options covered previously in the project.

Concurrent with the writing of the DM, a benchmarking study was initiated through the chemist on the team of competitor products in order to ascertain the current state of the market. In addition, a preliminary patent search was performed, with an initial set of qualifiers in an attempt to determine what design ideas have been patented in this product area. It was the NPI Team's intent to avoid lengthy studies on ideas that would violate existing patents as well as identify any designs that might be obvious extensions of existing ideas.

Benchmarking was one of the few formal tools of concurrent engineering used in this development process. As the case study shows, the benchmarking was not accomplished in a formal manner, since Hach's product was not directly compared to the competitor's, as true benchmarking would dictate.

As to other CE tools, the NPI protocol itself can be considered a concurrent engineering tool, since it represents a holistic planning for a development process. Theoretically, the NPI protocol gives the structure or outline to the mechanism to drive concurrency. In reality, it is up to the team to develop the informal team dynamics that make concurrency a reality or not.

Other than the benchmarking, the NPI schedule itself, and the Petri Net modeling, no other CE tools were used in this product development process. QFD is being considered for use in some product development projects, but the decision on which tools to use is generally left up to the team.

That QFD was not used does not mean that the "voice of the customer" was not heard in developing the product concept. The usual arguments are that product managers at most companies are familiar with their customers' needs and can speak for the customers. While this can be a very strong point, it must be realized that the product managers often speak using their perceptions of what the customers' wants are. This filtering of the customer's voice can alter the concept of a product measurably [6].

Hach Company has a very powerful tool that can be used by a product development team. One of the founding philosophies at Hach was that the education of customers using Hach instruments would be a keystone of the Hach culture. In line with this philosophy, Hach provides off-site training of users and maintains Hach Technical Training Centers (HTTCs) at both Loveland, Colorado, and Ames, Iowa. Classes in applications of Hach Instruments and Procedures run almost continuously at both sites for Hach Customers. These seminars, which usually run 1 or 2 days in length, give the customers hands-on experience with Hach Instruments in a laboratory setting after classroom introductions. Both in the classroom setting and in the labs, customers often discuss their specific applications and express their opinions on the instruments they use, both Hach's and its competitor's.

These classes are also open to Hach personnel, and NPI members are encouraged to attend. These personal interactions with customers are invaluable. From the designer's point of view, they offer a chance to ask questions and hear comments directly from users. Also, in a more esoteric sense, the NPI Team puts a face on the end user.

As an example of how this may change perceptions, a designer working in the R&D area may have the impression that the majority of the users for a certain instrument are the chemists in laboratories that he sees daily. Although the marketing representative may know that this is not the case, seeing in an HTTC seminar that the majority of participants are field users provides a different sense of the end customer and their needs. This can help eliminate some of the misconceptions that often spur certain features of a design.

Within 1 month, the NPI Team had developed three design concepts. Each idea, however, could be developed in two separate directions:

1. Using the current technology of the existing product; or
2. Incorporating a new technology developed in the basic research department.

The initial approach of the team was to thoroughly examine each option simultaneously, to narrow the focus of the NPI Team on only those ideas that showed promise of ultimate success. However, the NPI Team was persuaded consciously, as well as unconsciously, by members inside and outside the core team, to pursue only one concept. Conscious decisions in such a situation are driven by internal team discussions, based on individual experience and test results.

Often, unconscious decisions are made based on influences from outside of the team. A team that is relatively indecisive or internally undirected can be greatly affected by such influences. Much of this comes from the level of confidence a team has in itself. A team that is not sure it has the internal resources to handle more than a limited number of tasks can easily slip into a serial development process. A team that is confident of its technical abilities and decision-making capabilities has a great chance of success.

This confidence is difficult to develop, especially in a newly formed team. This confidence level is also affected, positively or not, by outside influences on the team. Low-level managers overriding team decisions can cause this confidence level to dip to the point of nonexistence. The experience of a knowledgeable manager outside of the team is an important asset to a project; however, when such a manager's influence, no matter how infrequently, is used to overturn team decisions on simple operational matters, the project can suffer. The view that a team has of management's confidence in the team is a great influence on how the team operates. Management providing clear objectives and support, focused on the strategic issues rather than applying their efforts to operational issues, is clearly key to successful team projects. Fortunately, the team described in this case study ultimately developed a much higher level of confidence in itself than it had in the early development stages. Also, as the team perceived that upper management was confident in the team's abilities, the decision-making dynamics of the group changed. This confidence lessened the impact of negative outside interference in later stages. The early decision to look at design concepts serially would ultimately cause the NPI Team to lose several valuable months of development. Once the 2 weeks of project strategy discussions were completed, work began on modeling the product concept.

The benchmarking study of competitor product was completed in 3 months. The results were circulated to the core team, and a meeting was held for discussion. At this time, the core team realized that a study of the current product was needed for comparison. Had the current product been included in the benchmark, duplication of effort would have been avoided, saving considerable amounts of time and effort by the chemist.

At the conclusion of testing of the initial product concept, the core team was holding biweekly meetings. The testing results indicated a need to dramatically modify the existing idea. Several members of the team were beginning to express some concern about not pursuing some of the other concepts, but, since the modifications to the initial design were not terribly time consuming, the issue of alternative ideas was dropped and the modifications were made. The testing of the modified design still was not promising, so the team focused very intensively over the next month on trying to make this idea work. Finally, after 2 months, the team gave up and went to the other ideas originally passed over.

Unknown to the core team, the third product idea had been studied for feasibility by one member of the core team and his subordinate. When the initial concept was scrapped and the other two product ideas were revisited by the core team, this member of the team reported that attempts to produce a model of the third were unsuccessful. Lengthy discussions by the core team did not provide any new concepts that met the project constraints, so the team was reduced to one idea. The project had proceeded for 5 months to date.

Although only one product concept existed, the forethought of one team member to independently pursue the third idea saved the core team from needing to study two additional concepts. The team was hoping to make up time by focusing all its efforts on the one remaining idea.

At the sixth month point in the project, the mechanical engineer (and project facilitator) gave notice to the team that he was leaving the company and returning to academia in 1 month. This was unexpected to the team, and the engineering manager assigned a new mechanical engineer to the project, to start in 2 weeks. Desperate to make up lost time, the team focused on functional aspects of the product design until the mechanical engineer (ME) could get caught up. Most of the work during the next 3 months was based on misconceptions due to poor team communications and assumptions by the ME that the current new product concept was well studied and firmly agreed upon by the team. The ME was new to the company and had only limited knowledge of how the NPI Team concept should be administrated. Much of the wasted time occurred because the ME trying to strictly follow traditional project management philosophy of Gantt charts, Pert charts, and so on, as he saw them being interpreted in the R&D area and not focusing on the design task at hand.

The ME, with input from the team, spent 2 months planning a schedule based on the interpretation of the NPI protocol as a serial process. As the plan was implemented, concept development continued at the original pace, but the actual progress in development and the plan diverged. The differences between the theoretical application of the NPI protocol to product development versus the practical application caused the team considerable frustration and eventually led the team to scrap the plan. The root problem, as seen from the view of the Facilitator, arose from the NPI protocol being serial in its strict interpretation and from the actual project being concurrent. Some see this as being the result of the "relative youth and inexperience" of the team at that point in the project. In any case, as communication improved, the strict adherence to NPI protocol by the ME was slowly modified into the more liberal interpretation used by several members of the team.

This change in the development process was slow, but, as the project neared the end of its ninth month, the design attempt at the second of the original product ideas was beginning to concern the team. By the eleventh month the second idea was abandoned. At year's end the team was, in reality, no closer to reaching their original goal than they had been near the start of the project.

At this point the project methodology clearly departed from the earlier serial approach. This led directly to three new product concepts developed almost entirely from within the team. The analysis, modeling, and proving of product concepts became distinctively concurrent. Within the remaining 6 months, the three product concepts merged into a single, well-defined, and manufacturable product.

One of the first tasks under the new project philosophy was to include additional functional groups of the company into the design process. Interfaces with Instrument Services, Ames Manufacturing, and HPP tool designers were established. The team analyzed the project as a whole by splitting out several components of the design. This carried the concept of project management one step further by creating sub-teams to manage the component subprojects. These mini-teams were basically autonomous, and critical review occurred only when the sub-team believed the component design was complete. In several cases, subprojects did not fulfill all the needs of the master project, and consensus was reached among the core team members to reevaluate that portion of the design. No subproject was considered complete until consensus of the core team was reached. Splitting out subprojects was not a conscious decision by the team; rather, it occurred as the result of the natural synergism between the appropriate members. Success of the subproject approach to project management was dependent upon the team's trust in it members. The knowledge that each member of the team would at some point be able to critique the work was what provided the necessary checks and balances to the process. The approach was used successfully on eight simultaneous subprojects.

It began to be apparent that the delineation of the strategic and operational decision boundaries was to be a key point in how this team operated within the corporate structure. Without analytical tools and using an NPI protocol that itself was in flux, it was not clear at times what the capabilities of the team were. A decision by the team to bring members from Ames to the corporate headquarters on short notice was a clear example of how such boundaries were defined at times throughout the project. Hach Company maintains corporate aircraft that are used to keep in touch with customers at their own sites, for business travel to other areas of the country, and also to transport Hach personnel between the different Hach sites. To use commercial air travel between Ames, Iowa, and Loveland, Colorado, is less cost effective in terms of time and money.

INTERFACILITY COMMUNICATIONS

In the recent past, new product development projects were contained almost entirely within a single facility, so the need for complete and timely project communication between facilities was not critical. However, as expertise has developed in different facilities and the company has restructured, this is no longer true. Many projects span two and even three facilities.

An NPI Team uses different methodologies in attempting to keep projects moving smoothly and efficiently. The different methods coincide with the change in development philosophy and corporate structure.

While projects were contained in a single facility, most communication occurred via monthly reports and sporadic telephone communication. The communication was mainly for information exchange and did not require the recipient to make decisions or take action. Therefore, this method of communication worked well. But the structure of the company changed, which started to draw other facilities into the new product development process.

For example, some projects started to integrate the Ames manufacturing facility into the project as experts in packaging and distribution. As new project teams needed advice on packaging regulations and distribution concerns, they now needed to communicate with another facility in another time zone. Formal communication to this point was still being handled mostly through monthly reports, which immediately proved to be to slow. Communication quickly switched from monthly reports to weekly reports sent through interoffice mail. However, there was still a several-day lag between an event occurring and its being communicated to the other facility.

Another problem presented itself at this transition. One NPI Team would not bring members from another facility into the product development cycle until the product was nearly complete in its design. Not surprising, projects were typically slowed as the other facility was now trying to fit their needs into a "completed" product. It was typical to see extraordinary modifications to existing processes to fit the new product. This could at times be costly and time consuming.

The next step was to have teleconferences that included members of the team from all facilities. This started the near real time communication between members in various locales. However, there are difficulties in communicating strictly by voice. NPI Team meetings were and are often used to display and discuss part geometry and fit. This all but demands visual as well as verbal communication. Facsimile transmissions became commonplace as this communication technique developed, but, at the height of the development cycle, parts would change so quickly that drawings could not be sent fast enough and it was very difficult to identify problems with parts and assemblies fitting together.

Parts and assemblies were mailed between facilities, but (again because of the time lag) there was still some delay in getting processes in other facilities started until part geometry was completed and set. If a cartridge needed to be filled, it was very difficult for production engineers to design the machinery needed to fill the cartridge until they had cartridges. Drawings and telephone conversations were no replacement.

Finally, there were two technologies incorporated into the company that have increased the efficiency of the development process: the local area network (LAN) and video conferencing.

The ability to exchange drawings, minutes, and other files directly through the computer system was invaluable. What this means is that any member of the team can access files that are as current as possible. If a meeting is scheduled, the most current information can be viewed just minutes before the meeting. This keeps the entire team up to date, regardless of location. This is a tool that should not be overlooked in its importance to communication.

Video conferencing is the ability to communicate live with multiple locations simultaneously in both picture and sound. It is the next best thing to having all members assembled in the same room. It has minimized the need for travel between facilities, which is costly and time consuming. Also, it lets every person at every facility see what the parts look like and how they fit together and, in many cases, allows a considerable amount of "mental" design and experimentation to occur long before actual parts are available.

With all of the new communication tools, every member of the team can be involved face to face through every step of the product development cycle with the most up-to-date information available. Even so, at times true face-to-face communications are necessary between different areas.

An NPI Team wanted to have a representative from Ames Manufacturing and the member from Ames Service out on short notice at a key point in the project. The team facilitator was not sure whether such travel costs would be carried by the project, R&D general accounts, or the departments that the members came from. Once this question was resolved after lengthy discussions, written justification had to be developed and sent to one of the individual managers involved in Ames before arrangements could be set. At the time this appeared to be necessary, yet the efforts involved could have been lessened if it was understood that the team was capable of and had the prerogative to make clear, well-founded decisions on such problems.

Even with clearly defined boundaries, it is often necessary for a team to discover where its boundaries exist. In this project it was found that a few basic questions could be applied to solving boundary dilemmas. Is it spelled out that we can do it? Is it clearly defined that we cannot? Do we have good reason for taking on the responsibility? This is simple analysis, but one that proves to be accurate in most cases. Unfortunately, such analysis can lead to conflicts with those who have different opinions on the answers to the questions.

Also, once the change in project philosophy was fully entrenched, several misunderstandings occurred between the team and management. Analysis of the situation revealed that the rate of change to the product outpaced the rate of formal communication to management, often times causing management to appear out of touch with the day-to-day functioning of the team. Where the team and the particular manager agreed on the operational boundaries of the group, this did not cause conflict. In some instances and from some viewpoints, the team was seen to overstep its decision-making bounds. This led to conflicts that took effort from both the team and lower-level management to resolve.

Increased and more timely communication between the members of the NPI Team is obviously one of the major linchpin pins in using concurrent engineering as an NPI tool. As part of a general continuous improvement philosophy throughout Hach, the NPI Team recognized that a study of the communications—both within the NPI Team and to different areas of Hach—could be beneficial. Throughout Hach, traditional flowcharting is used to study how other groups accomplish tasks, as a tool to improve systems. Flowcharting is a valuable aid in studying concurrent engineering, since concurrence can be mimicked somewhat by studying narrowly bounded areas of a process in small steps. However, while flowcharting is entirely adequate for serial operations and limited concurrent processes, the operations of this NPI Team could best be studied using engineering tools for concurrent processes.

As chance had it, at this point in the project the team facilitator was completing course work at Colorado State University for an MS degree in Mechanical Engineering. By allowing employees to work flexible schedules and providing financial incentives, Hach Company encourages continual formal education for all employees. During a graduate course in "algorithmic control," which covered a variety of development schemes for software and the control of machine operations, the discussion turned to the Petri Net.

The thought process that led to the development of the Petri Net model of NPI Team communications was driven by a member being in a classroom listening to solutions to problems being described. The fact that the problems being solved were not directly related to those that person had in mind was probably immaterial.

It is difficult to disassociate entirely from work-related problems when the classroom session is in the middle of the day and the problems are serious. In this case the natural link between the NPI problems and the Petri Net solution were immediately evident. The synergy that is hoped for between academia and industry on even a small scale became a reality in this case.

This study of communications showed natural boundaries of the strategic and operational planning areas. Of course, such a model may not be perfect due to the large number of communications moving among even a small group. It does give a feel for where such boundaries may naturally occur or how they can be drawn to optimize the development process. A last caveat is that the method, as with any such model, is somewhat open to individual interpretation. When used as an analytical tool, the effects of interpretation are minimized.

RULES APPLIED TO THE PETRI NET MODEL

A Simplified Timed Petri Net model and protocol was developed. The study of this model, as shown below, led to a number of conclusions about the communications of the NPI Team. Based on these conclusions, improvements were made to the communications process, leading to an improvement in the development process itself. The Petri Net model for this problem is shown in Figure 4-3.

The rules for the Hach communications study are described below.

1. The study is of the routine communication of an event from the NPI Team from the view of the facilitator and covers only the Loveland Plant.
2. The study covers formal written communications and informal meetings.
3. The "layers" in the communications are:
   a. New Product Implementation Team (NPI Team).
   b. Support Group.
   c. The immediate engineering supervisor.
   d. Vice president, Product Development (VP Prod Dev).
   e. Senior vice president, Research and Development (Sr. VP R&D).
   f. Technical Committee (upper management).
4. Assumptions are:
   a. No one person is missing for over 1 week.
   b. All levels must be covered.
   c. Individual team members will communicate within their own areas outside of this study.
   d. Monthly reports take 1 week to write and are distributed 4 weeks after the event occurs.
   e. Meeting minutes take 3 days to write and distribute.
   f. Biweekly objectives take 2 days to write and distribute, 1 week after the event occurs.
   g. Formal communications occur concurrently along different paths.
   h. Informal communications can be serial or parallel.

84   Successful Implementation of Concurrent Engineering

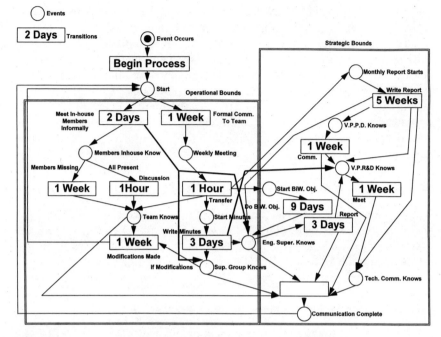

FIGURE 4-3.   Petri Net Model developed for this case study problem.

5. Formal communications use the formats now in place:
    a. Weekly meeting minutes.
    b. Biweekly objectives.
    c. Monthly reports.
6. Informal meetings are 0.5 hours in length, and covering the complete NPI Team (ten members) requires the full 5 hours.
7. The event being communicated happens the first day of a month. This should be worst case (just after a monthly report is completed).

WHAT THE STUDY SHOWED

The model was studied by observing the flow through the communication process using markers. The time for the event to be known at each of 8 points was tabulated. Table 4-1 illustrates four pieces of information for the communication process. These pieces of information are:

1. The quickest path, using the original method.
2. The quickest path, using a modified method. This method was the addition

TABLE 4-1. Model Observation of Timed Flow Through the Communication Process

	Quick Path, Traditional (Days)	Quick Path, Modified (Days)	Slow Path (Days)	Redundancies
NPI Team knows	2	2	5	0
Support Group knows	8	2	8	0
Engineering supervisor knows	8	2	8	2
Entire operations 1 group knows	8	2	8	2
VP Prod. Dev. knows	11	11	30 (6 weeks)	1
VP R&D knows	18	8	30 (6 weeks)	2
Technical Committee knows	25	15	30 (6 weeks)	1
All know	25	15	30 (6 weeks)	8

of informal communications to the Support Group and weekly minutes being sent to the Sr. VP of R&D.
3. The slowest path, using the original method.
4. Redundancies—The number of extra times the same information was communicated.

The formal communications do produce redundancies. Assuming that everyone reports most of what they do on a regular basis through the NPI Team, the biweekly objectives seem to be the one communication that could be eliminated. This would eliminate two redundancies. The other redundancies come from the assumption that each level passes on information that has already been communicated through other formats. Possibly, some of these could be eliminated.

The first of the two suggested "new routes" (shown by bold lines) increased informal communication with the support group. This was to get the support group more in tune with what is happening with the NPI Team. The second route, which consists of copying weekly minutes to the Sr. VP of R&D, was at his request as the project was moving into new areas. In this case, upper management had clearly empowered other levels so that the communications were truly for information only and did not effect operational decision making. At least in getting the information to this level, the time savings were relatively high (10 to 14 days). Also, improvements were made to limit the communication time of formal written communications between the Hach

86   Successful Implementation of Concurrent Engineering

Facilities. While Hach currently does not use an E-mail system, a pseudo-system was implemented. This consisted of the creation of a common directory on the Loveland Local Area Network Server, which is accessible by all team members (or other interested parties). This shortened the path of written communications by allowing the bypassing of intercompany mail. While the mail system is fast, considering the physical separation of the facilities, at best it still builds in hours of slack time to these communications. Also, the team started using teleconferencing with its members in Ames. While this was a decided improvement over strictly written communications, it was limited by the lack of visual contact. Recently, Hach has implemented video conferencing between the facilities, which is another major step forward in lowering communication barriers. While physical visits are still necessary at key times in the process, the use of these methods is key to shortening the development times of products being developed in facilities that are physically isolated from each other.

On a larger scale, the study reinforced the generally accepted view of how concurrent engineering can function at its best if the "lower levels" (upper levels in an inverted pyramid structure) are empowered to act on events as they occur, leaving management to function as a support and general guidance mechanism. Figure 4-4 shows the increase in reaction time to an event based on the level that the communication of that event has reached.

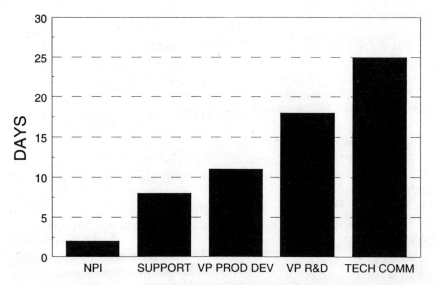

FIGURE 4-4.   The increase in reaction time to an event, based on the level that the communication of that event has reached.

If a decision is made by the NPI Team, it can be implemented in 2 days. If the decision were to be made at (as an example) the VP level, no movement in the process would be apparent for a minimum of 11 working days. Fortunately, this was not the case throughout this project.

Simply, for a company to successfully implement concurrent engineering, it is imperative for the company to understand its communication mechanisms. A team that is making design modifications on a daily or even hourly basis requires a system of communication and feedback that is equally as responsive as the concurrent design process.

References
1. Murata, T. 1989. Petri nets: Properties, analysis and applications. *IEEE*, 77(4):541–561, April. Invited paper.
2. Yamalidou, E. C., and J. C. Kantor. 1991. Modeling and optimal control of discrete-event chemical processes using Petri Nets. *Computers in Chemical Engineering*, 15(7):503–519.
3. Bosch, M., and G. Schmid. 1991. Generic Petri Net models of protocol mechanisms in communication systems. *Computer Communications*, 14(3):143–156, April.
4. Hillion, H. P., and J. M. Proth. Using timed Petri Nets for the scheduling of job shop systems. *Engineering Costs and Production Economics*, 17:149–154.
5. Hatono, I., K. Yamagata, and H. Tamura. 1991. Modeling and on-line scheduling of flexible manufacturing systems using stochastic Petri Nets. *IEEE Transactions on Software Engineering*, 17(2):126–132, February.
6. Design it right, a tale of three multimeters. EDN, October 1, 1992.

5
Concurrent Engineering at Chipcom: Implementation at a Small Corporation

John McNamara
Chipcom Corporation

INTRODUCTION

The purpose of this chapter is to highlight specific issues that Chipcom Corporation faced when implementing concurrent engineering. Chipcom is still in the early stages of execution and continues to pursue solutions to issues raised by changing the company's approach to new product development. Some choices were contrary to accepted concurrent engineering wisdom, such as the method of implementation, but were correct for Chipcom, given its culture.

During the last quarter of 1991, a cross-functional team of employees, representing all parts of the organization, began working on a plan to identify opportunities for improving its new product development process. As a small, growing corporation, Chipcom faced two conflicting problems: a lack of resources and rapid growth. The drive to implement concurrent engineering techniques presented Chipcom with an opportunity to leverage its resources and develop a solution to this dilemma. The team began with the intention of defining the boundaries of concurrent engineering, but realized that these methods had broad implications for the entire company. The changes the team planned would cut across all parts of the organization. The acronym PACT symbolized Chipcom's application of concurrent engineering, at the highest level. In other areas, this is referred to in terms such as *strategic product design*[7] or *integrated product development*[5]. The group evaluated many definitions of concurrent engineering, but felt that each lacked the specifics that would identify it with the values and practices of Chipcom. It was essential that any statement that defined concurrent engineering stay rooted in the Chipcom culture and build on the organization's systems and procedures.

PACT stands for Partnerships for Achieving Champion Teamwork. It is defined as "a model for new product development incorporating customer needs and early supplier involvement with cross-functional teams to improve time to market, 'do it right the first time,' optimize the use of resources, and continuously improve customer satisfaction."[3] The term PACT also signified the agreements made within the organization to bring a project from product idea to the marketplace, based on an agreed-upon business plan.

THE CORPORATION

Chipcom is a small company, founded in 1983 to design, manufacture, and market fault-tolerant, intelligent switching hubs. "The Chipcom product lines focus on managing dissimilar computer products within a building or group of buildings. Beginning with the first product line, the Ethermodem family, to today's family of ONline System Concentrator products, Chipcom has focused on delivering products that supported enterprise-wide computer networks termed Facility Networks."[2]

In 1991, ..."Chipcom's revenue was approximately $50M. This was an increase of 69% over the previous year. The company also introduced 20 new products, the majority supporting the ONline product family."[1] Chipcom sells its products through value-added resellers and system integrators. The company is headquartered in Massachusetts, with sales and support offices throughout North America. Chipcom Europe BV is an independent entity that focuses on marketing, sales, and support through resellers in Europe, the Middle East, and Africa, with subsidiaries in France, Germany, and the UK. The company also has engineering offices in Israel.

The Corporate Management Team (CMT) leads this corporation, organized along functional lines. The CMT consists of the founder and chairman of the corporation, the president and CEO, the vice presidents of Marketing, Field Operations, Engineering, Finance, and Human Resources, and the director of Manufacturing.

THE CHIPCOM SMARTHUB

Since the company's facility network solutions provide high reliability, a natural move was to incorporate fault-tolerant technology into its products. The Chipcom ONline System concentrator provides a modular, fault-tolerant platform for use in facility networks. The product was unique because it could handle up to three simultaneous Ethernet, Token Ring, or FDDI networks in a concentrator, with the capacity for bridging and routing. Few products are capable of linking different LANs and media types within the same platform. Auto-switching power supplies and redundant links through automatic switch-over, combined with backup technology, delivers fault tolerance to the ONline concentrator.

CHIPCOM CULTURE

Chipcom has a culture that is team focused. The Project Management Team, in which representatives from each functional group work to bring a product from concept to market, is an example of the team approach. To ensure that a person is "chipper," or fits into the team atmosphere, prospective new hires get screened by representatives from various groups within the company. Teams form regularly to complete specific tasks and then dissolve when the tasks are complete. During employee orientation, each new employee receives a copy of the company's Value Statement, which covers four areas: pride, quality, respect, and sound business practices. While it is displayed in the company lobby, it is also not uncommon to find a copy of the Chipcom Values hanging in any employee's office.

THE CURRENT NEW PRODUCT DEVELOPMENT PROCESS

Since 1987, the company has used a documented procedure to develop new products. The New Product Development Process, or NPDP, is a major contributor to the success of the corporation. Its dominant feature is a process containing six sequential phases, with a phase review (necessary for transition from one phase to the next). A small, cross-functional core team called the Project Management Team (PMT) is given the responsibility of developing, marketing, and introducing a new product to the market. The PMT is also responsible for specifying the deliverables of each PMT member, developing checklists to assist each PMT in ensuring that all items of a new product are considered, and, finally, attending an informal monthly review, as required, to highlight any actions that require the CMT's assistance to ensure the rapid delivery of new products to the market.

The phase definitions are as follows:

Phase 1—Investigation
Marketing and Engineering investigates product ideas to identify those that warrant further development.
Phase 2—Definition
A product and marketing summary defines the product requirements and technical product definition and identifies a proposed project plan. Approval by the CMT is needed for the project to go forward.

Phase 3—Design
The hardware and software engineers complete the actual design of the product, while Product Engineering evaluates whether the product meets design goals and performance requirements. Peer design reviews ensure that engineers use the best design techniques.

Phase 4—Verification
Product Engineering verifies through the preproduction build and controlled hardware and software testing that the document package and the manufacturing process are complete. Regulatory product testing ensures that the product meets the guidelines established by the FCC, BABT, and other regulatory agencies.

Phase 5—Introduction
Manufacturing produces the first production units, using the supplied documentation package. Upon completion, the team reviews the product preproduction results for approval of first customer ship.

Phase 6—Production
Full-scale production is underway. The team holds a 3- and 6-month product review to highlight any product or process problems and to identify resolutions to the problems.

These sequential steps outline a fairly traditional NPDP (Fig. 5-1), with the unique feature of the PMT. It is the focal point of the process and links all the resources of the corporation into a cohesive group responsible for bringing the new product from definition to first customer ship. There are other elements of the NPDP that prove beneficial to the PMT: The limited parallelism in the process helped to minimize cycle time, and, more importantly, a sample project schedule provided a model to assist the PMT in developing project plans. This gave each PMT a common language, process, and understanding of what it would take to successfully introduce a product into the marketplace. Each PMT is formed for the purpose of delivering the one product and is dissolved at the completion of this task. The PMT consists of representatives from Marketing, Tech Support, Engineering, Advanced Manufacturing Engineering, and Product Engineering, giving the team the resources and responsibility for the entire project. "The PMT has overall responsibility for the project. If a project succeeds, the team takes credit; if it fails, the team is accountable. It is important for all team members to be involved, even during phases when their roles are minor. It is not necessary for each member to attend every meeting or understand every issue, but it is important for PMT members to know generally what is happening, and if there are problems that are not being solved to escalate them."[6]

The strength of this process is its simplicity, its documented roles for each functional group, checklists for ensuring that all tasks are complete, a sample

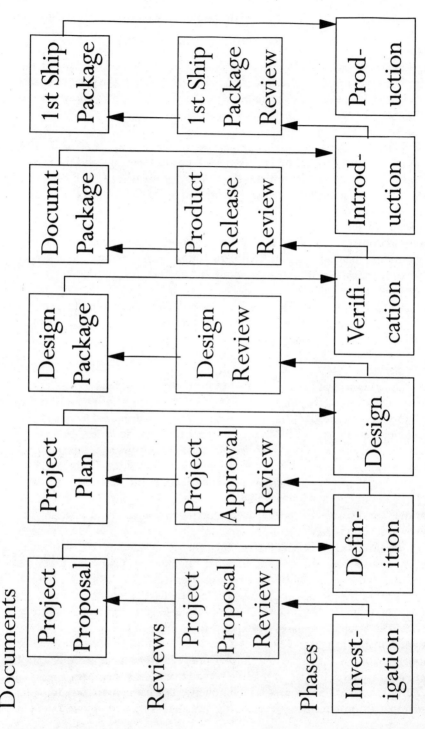

FIGURE 5-1. New Product Development Process (NPDP).

project plan, an identified review cycle, a list of deliverables, and the creation of a focal point for new product development—the PMT.

Some of the process strengths were also its weaknesses. The documented process resulted in resistance to process changes demanded by the needs of each project. NPDP began as a total to facilitate the successful completion of projects, but, as it became incorporated into the culture, people were less willing to deviate, even when the nature of the product suggested changes to the process. The temporary nature of PMTs was another weakness, since those that performed well disbanded at the end of their project. PMT evaluation meetings, which evaluate the performance of the team, are a method for passing along information concerning the success or failures of the PMT. Other weaknesses resulted as the needs of the organization changed. Since the focus of the project shifted from one functional group to another, the responsibility for the project's completion became blurred. It is difficult for the CMT to identify resource requirements as the number of new projects grows. Tools that monitor the overall resources of the company are still cumbersome, but are being improved. Project management software that can track resources and link multiple project plans together, summarizing the results, is essential for the company in determining if engineering resources are available for a new project.

WHY CHANGE?

Chipcom has a very positive view of itself, with continuous, healthy growth year after year. Top industry analysts viewed the company as a winner, with a strong product line and presence in the SMARTHUB market. The company had gone public in 1991 and then followed up with an alliance with IBM in 1992. Considering Chipcom's accomplishments, the company could have concluded that there was no reason to change. There were several issues, however, that forced many in the company to reevaluate the corporation's performance and to conclude that the things Chipcom did to achieve its current success would not necessarily ensure future success.

First, there were external factors that Chipcom would have to respond to, but over which they had little direct control. Product life cycles declined, as technology advanced at a fast pace, allowing the addition of new features to enhance product functionality. In contrast to this, customer sophistication and experience grew, resulting in a greater demand for product variety and very competitive pricing. Customers were pushing vendors to make their products interoperable. This means that customers have less reliance on any single vendor and that corporations such as Chipcom would have a greater need for the latest technology when the customer needs it. Customers need quick delivery of product, since they are also trying to meet the immediate demands of their customers.

Lastly, there were internal factors over which Chipcom had greater control to modify or change. Chipcom's market position is related directly to customers' and resellers' perception, not on its actual market standing. Chipcom's performance measures were not quantitative enough and did not target specific areas for improvement. The company was still operating with methods that worked for a smaller company, and rapid growth had put strains on the internal processes. Most importantly, Chipcom was competing with companies that had more resources, market share, and people and was intent on overtaking those competitors.

The Chipcom culture is compatible with the changes that concurrent engineering proposed. Concurrent engineering was a method of improving Chipcom's new product development process through a framework that effectively identified and delivered customer expectations, achieved improved quality through consideration of problems early in the design cycle, optimized the cost and performance, and delivered the product to market faster. It is not the simplistic idea of reducing the time allowed for each step; rather, it is the effective use of Chipcom's skills, communication ability, and technology. It is doing things smarter and faster, not simply faster. Concurrent engineering made sense as a method of solving the problems presented by both internal and external factors.

CONCURRENT ENGINEERING'S EARLY STAGES

During the summer of 1991, the CMT began to reevaluate the methods that it used to bring new products to market. The reasons were obvious to the CMT. First, it recognized that new products were fueling its expansion into the new SMARTHUB market and that a constant supply would be of greater importance as Chipcom expanded outside the Ethernet market into other protocols, such as Token Ring and FDDI. Furthermore, the number of products needed to establish Chipcom as a major player in the SMARTHUB market required a faster new product development cycle time. Finally, a conservative approach to business practices meant controlling expenditures and especially new hires so that headcount would not grow faster than the business could support. Improving the new product development cycle time provided a solution to these issues.

Mid-July was the beginning of planning for 1992, and resources were a major consideration. Most departments (but, as in particular, Engineering and Manufacturing) saw concurrent engineering as being an important contributor towards accomplishing the product goals in the following year.

During 1991, the company planned a 51% increase in headcount, but the coming year would not be so aggressive. Clearly, it was vital to look at concurrent engineering for its ability to bring together teams within the company to speed product design and improve product quality.

Discussion about the value of CE to Chipcom and how it could improve the new product development process was very active during the summer of 1991. The vice president of Engineering distributed an E-mail that outlined some of his thoughts and a document that detailed a working definition, a plan for implementing changes to the NPDP, and the culture changes required by the new process to key people in the organization. Strategy meetings continued to highlight the need to further evaluate and study the benefits of concurrent engineering. In September 1991, the vice president of Manufacturing assumed the responsibility of determining what concurrent engineering would mean to Chipcom and how best to take advantage of it. Chipcom hired a consultant to focus the company's efforts and push the process along when it got stuck. This led to the first meeting, in October 1991, that specifically addressed the subject of concurrent engineering as it pertained to the NPDP.

October 23, 1991

The October 23 meeting brought together about 30 key people from all parts of the organization to have an open discussion about what shape concurrent engineering would take, what the term meant, and how it could specifically improve the company's existing process. The meeting was run by a consultant, who facilitated discussion and served to keep the discussion moving forward. In addition to key department representatives, the CMT played a major role in the meeting and the vice presidents of Engineering, Marketing, and Manufacturing attended the all-day session. Each participant had varying levels of knowledge before attending the session about concurrent engineering and how Chipcom might apply it to the NPDP.

The meeting began with a presentation by the vice president of Engineering, who reviewed the company's new product development performance for the past 3 years, as well as the changes in product development cycle time. His belief was that, to continue to be competitive, the company must reduce the product development cycle time aggressively. Performance showed consistent improvement, but, without a mechanism to continually improve the process, he felt that PMTs would begin to stagnate. A revitalized approach was essential. He proposed the following challenge to the assembled group. Should concurrent engineering be Chipcom's solution to continued improvement? The goal for the group was to discuss issues related to product development and Chipcom and to select a task team. The task team's responsibility would be to take the issues highlighted by the group, develop an implementation plan that would fit within Chipcom's culture, present this plan to the CMT for further review, and work with the CMT to develop for Chipcom a process for developing new products. This goal was developed over time and was not evident at the time of the meeting. The only thing clear at the time of the meeting was that people were unclear as to what concurrent engineering was, in relation to Chipcom.

The consultant began by leading the group in a discussion about concurrent engineering through a tool called an *affinity diagram*. In this case, the participants constructed the affinity diagram with Post-it note's that contained up to a three-word phrase that described each person's view of concurrent engineering. The participants could use no more than five Post-its and were instructed to make no assumptions, but to "live within Chipcom." In other words, the ideas should make sense within the culture at Chipcom. The notes were then placed on the wall and in silence all participants rearranged the Post-its into categories and assigned headings. The group generated about 150 ideas in 20 minutes and organized them as the basis of discussion for the day. The top five categories from the discussion in order of percentage responses were:

1. Time to market   15.1% of Post-its
2. Team process   13.5% of Post-its
3. Vision   11.4% of Post-its
4. Respect   11.0% of Post-its
5. Customer   10.6% of Post-its

Issues revolving around the product development process and specific actions that could improve the current processes were also a focus of discussion. Specifically, the discussion focused on the role of the PMT and leadership.

The PMT is an important part of the Chipcom NPDP. It is a team containing members of each functional group having a stake in the product design that forms for the purpose of bringing a product from concept to market. The PMT is a strong part of the culture and fits nicely into the concurrent engineering concept because of its inherent focus on teamwork. The shortcoming of the existing process, from the viewpoint of many members of the gathered group, was its focus on six sequential steps that passed product responsibility from one member to the next. The challenge would be to keep the best parts of the current NPDP and integrate them into the new concurrent engineering process. Several themes repeated themselves during the day's discussion that highlighted aspects of importance to a successful process at Chipcom. These themes are as follows:

Ownership of the product—PMTs must show by action that they have a commitment to getting a product to market by the date that they agree to. Actively lobbying within the company for resources, making management aware of ways to improve the schedule, and keeping team members informed of progress through constant daily communication contribute to a sense product ownership.

Dedication to the customer—The sense that the customer needs the product and that the team has a responsibility to get the product to market to meet customer needs. This is a close cousin to ownership of the product.

Project management skills—A basic skill needed by all members of the team is project management skills. Training and project management tools were essential in effectively managing the product delivery schedule, particularly in the end game.

Team membership and leadership—This subject elicited the most discussion because of the natural conflict between leaders named to their position and leaders within the PMT recognized because of their knowledge or past contributions or by virtue of their strong personality. Should the team process be a single group for the life of the product, or should members transfer in and out of the team as needed? Should team leadership change as the focus of the product development changes? Should the team leader have the authority to select members of the PMT? These were all questions that were debated vigorously.

Meeting Conclusion

At the end of the day's discussion and with a greater sense of the issues that face Chipcom, the group selected representatives from each group to form a concurrent engineering task team. The team consisted of members from Product, Hardware, Software, Advanced Manufacturing Engineering, Materials, Quality, Finance, and Marketing. It was responsible for developing a plan to present to the CMT that would outline the investment required by the company to implement concurrent engineering. The group needed to be as specific as possible in detailing a timetable for implementation and identifying the necessary investment by quarter.

THE PLAN

The task team had its first meeting a few days later to establish the goals for the team and to settle on some logistical questions. Three additional team members joined the task team from the original eight chosen at the October 23 meeting: Management Information Systems (MIS), Technical Support, and Human Resources. The team selected the HR representative as the task team leader. It felt that HR was the most neutral member and could be the best arbiter of any conflicts between groups. It also set 6 weeks as the target delivery of a plan to CMT. To do this, the team determined that they would have to meet twice per week for 3–4 hours, with one full-day, off-site meeting to finalize plan details, leaving the last week for writing and developing the presentation.

Some of the issues that the group determined were essential to address were the form that concurrent engineering would take, the method of implementation, the required investment, and how Chipcom would measure return on investment. At the formation of the task team, a working definition of concurrent engineering had not been developed, but the team did know the elements that should be incorporated: customer needs, time to market, cross-functional teams, supplier involvement, and continuous improvement. These were the topics identified as important during the off-site meeting. The team agreed that the plan should take the form of a business plan, since the organization and topics covered would be familiar to the members of the CMT. It was also recognized that the target audience for the proposal was split into two distinct groups. First, there would be members of the CMT, who would view the proposal in terms of an opportunity with potential risks and rewards to the company, and its long-reaching effects on the Chipcom culture. Second would be the employees of the company, who would tend to focus more on what particular effect the changes would have on their position and responsibilities.

The PACT Process

The PACT plan converts the existing six sequential phases into three phases: investigation, definition, and development. The first two phases would still remain sequential, with CMT review and approval after each step, while the development activities occur in parallel and do not have the phase review component of the previous process. The development phase is not entirely concurrent; there are some activities that have to occur sequentially. Printed circuit board layout cannot occur in parallel to printed circuit board design, as an example. Discussions with the CMT highlighted the need to get early feedback on the product ideas and its marker position. The purpose of early CMT involvement was so that they could have the greatest opportunity to influence the product before the team completes the business plan. The PMT is organized earlier, after the investigation phase, in order to give each PMT member the opportunity to formulate the product definition and negotiate the features necessary for the development of the *product business plan*. It was an important consideration of CMT and PACT to limit the documentation to an amount that would provide an "adequate" business plan. If too much time is invested in completing the Investigation and Definition stages, it becomes a barrier to entry and only those products that are "sure things" are worked on.

Business Opportunity

The front end of product development can be very difficult. A new product idea is fuzzy and has many uncertainties in the early stages. There is no funding to work on the new product and few people are aware of it, so considerable time is spent simply explaining the new proposal. Without a strong champion, the idea can easily be delayed while satisfying other priorities, with a significant amount of time being lost.

The organization also has a need for information about a new product so that it can differentiate between an idea that is a good strategic fit and one that isn't. Chipcom wanted to establish a clear beginning to the process and create a method of capturing the "underground" projects. The Business Opportunity Statement is a 1–2 page document that describes the customer need, the product that addresses that need, the market, the fit with Chipcom's strategy, and business potential.

Investigation

The investigation phase focuses on customer needs, how a product concept fills those needs, how it fits into the corporation's strategy, as well as revenue projections, risks, obstacles, and the product importance to Chipcom. This is the point where tools that uncover customer needs, such as quality functional deployment (QFD), are used effectively. The investigative phase is carried out by a product and an engineering manager, who collect and assemble the information into *market requirements*. The plan is presented to the CMT for recommendations and approval to form a PMT and to get the next round of funding for the definition phase.

Definition

Once the CMT gives approval to form a PMT, the task at hand is to deliver a business plan for the product. The product is not simply the physical hardware and software, but also the technical support that the organization supplies, the educational training, and the documentation for the product. The business plan consists of specific product requirements negotiated between the PMT members. Trade-offs satisfy the needs of all representatives on the PMT; thus, the product is defined by features that optimize the needs of the company. The plan also includes further explanation of the marketing opportunity, financials, detailed manufacturing, technical support, sales plans, and a detailed project schedule. The deliverables from this phase are required for full allocation of funds for the project by the CMT. Assuming there are no changes in the market conditions or significant deviations from the product business plan, there are no additional, formal phase reviews required by the CMT. The assumption is that, if the PMT is following the business plan, the process does not need to be delayed as the team stops and prepares for a phase review.

Development

The former design, verification, introduction, and production phases of the NPD process are incorporated into the new development phase. This is because these activities can happen in a concurrent manner accounting for interdependencies, while those activities that are independent occur in parallel. Peer reviews as well as technical reviews by the functional groups are used, but formal CMT reviews are unnecessary. This phase would make use of electrical, thermal, and process simulations prior to committing to a hard prototype. Since there was considerable planning up front and a detailed description of the product requirement, all members of the team have the same information about what the product will be. The team needs to invite suppliers into the PMT meeting, to assist the team in manufacturability issues. Since design engineers and manufacturing engineers are familiar with only a small subset of components and processes, the suppliers bring a greater variety of experience and choices to the team. As part of the team and by being involved at the beginning of the design stage, the suppliers can provide realistic information concerning deliveries of parts and can buy into the project commitments.

PACT and CMT Leadership

The CMT played a critical part in refining the PACT details and in adding critical pieces to the process. Achieving the level of change that PACT presented would have been impossible without the involvement and commitment of CMT. The consultant who worked with the PACT task team termed this commitment as "walk the talk." This commitment was demonstrated by the amount of time that the CMT scheduled with the PACT task team, giving the team the opportunity for continued discussion well beyond the initial presentation; by providing process visibility and support in the organization, through presentations to groups within the company; by funding necessary changes, such as QFD training, off-site meetings for the task team, and other subgroups, and by hiring people focused on employee training. Finally, the

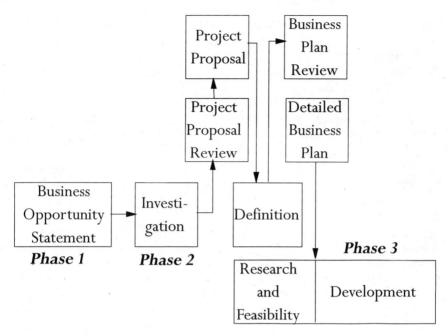

FIGURE 5-2. PACT new product development process.

VP of Human Resources was assigned by the president to drive the process forward in the organization. There were several 4-hour sessions in which the CMT and the task team discussed, modified, and improved the plan. Several small groups were formed that solved, with the help of the CMT, specific issues, such as the required information to move from investigation to definition phase and key points for CMT review.

PACT Implementation

The task team looked at the possible methods of implementing this plan and felt that the number of changes to the organization and the amount of training necessary was too high to accomplish on every new product started. This led the team to recommend a pilot program with three to four products that would follow the PACT process. This has to be preceded by training for the PMT members in project management skills, team building, and QFD. The NPD process would have to be rewritten so that each PMT knew the expectations and process. The CMT viewed this plan as having several drawbacks. It would mean a delay of a year before all products switched over to a new process, and the organization would have to wait before realizing benefits. It meant that during that time there would be two different processes, creating considerable potential for confusion. It also meant that the PACT process needed debugging before being introduced to the company. This created concern among the team that a process that did not work would set back the efforts of the team in introducing change to the new product development process.

CMT recommended not to pilot the plan, but to implement individual pieces of the plan. There also would be a risk of confusion in this approach, but there would be positive results sooner. It was felt that the risk to the company in waiting was greater than in risking some confusion. The CMT also expressed confidence in the organization to adapt to the changes and felt that the organization needed to see some positive changes.

The PMT

In the PACT proposal, the PMT is made up of representatives from Marketing, Development Engineering, Product Engineering, Advance Manufacturing Engineering, Technical Support, and key suppliers. The team is a group of strong professionals who own the responsibility of delivering a new product to the market. Each member has a responsibility to represent his or her functional organization, as well as to make sure the PMT gets the answers and resources it needs from the different organizations, to keep the project moving forward. Additional responsibilities of the PMT members are still undefined, but a clear definition of the goals and objectives of each representative were recognized by the PACT task team as essential to a smoothly operating team.

PMT members are peers, with no one member having decision-making authority over the others. This feature of the PMT can cause conflicts, so a respect for each member is important, as is the willingness to listen to each member's opinions. A unanimous vote will allow the PMT to move forward, but objections will stop the process. In that event, the member with the objection will take action to resolve the conflict so that the team can work back towards establishing a consensus. Those issues unresolved will be escalated to the PMT manager (see "The PMT Manager").

The PMTs have a room in the company set aside just for team meetings. The room is bright and is equipped with a conference phone and a white board. It is comfortable and conducive to holding discussions without interruption. Meeting times for each approved PMT are published, so everyone knows when a particular product meeting is held. The procedures for running meetings are known and well established due to the documented new product development process. Team building skills are at the heart of the PMT and viewed as being essential to the PMT's success.

The PMT Leader

Leadership of the PMT was a major topic of discussion, both among the PACT task team and the CMT. Strong leadership was recognized as a key factor in successfully getting any product to market quickly. Leadership is not a skill that all members of the PMT would have; therefore, the leader should not necessarily come from any particular functional group. The most important feature of the PMT leader was that he or she would be someone who was capable of effecting change and who would know when to force the PMT to come to a decision and when to escalate a problem. It is an administrative role performed by one of the PMT members who calls meetings, sets the agenda, keeps track of the action items, publishes minutes, makes sure that all members of the PMT have the opportunity to contribute, and escalate issues to the PMT leader in the event that the PMT cannot come to a consensus. In practice, the PMT leader is chosen by the PMT itself. If the PMT neglects to appoint a leader or cannot come to agreement, the PMT manager will appoint the leader.

The PMT Manager

The PMT manager role grew from Chipcom's need to get decisions made quickly. Since the corporation was growing quickly, many of the people who had more experience with the corporation and who were more familiar with the strategic direction moved up in the organization, while newer employees took their place. This meant that the PMTs had less people who were knowledgeable of Chipcom and who fully understood the company's strategy or the interdependencies of an increasingly complex product.

In the early days of the company, access to the CMT was very easy. Decisions that the PMTs couldn't resolve because there was no consensus were quickly escalated to the CMT and solved. As the corporation grew, the responsibilities of the CMT changed from internal day-to-day product issues to external issues, such as customer visits, press meetings, discussions with analysts, and so on. PMTs increasingly found it more difficult to relate individual product decisions to the corporation's priorities. The result was that decisions did not get escalated, because newer employees were less comfortable using the CMT as a tiebreaker, or else those PMTs that did escalate problems found it more difficult to schedule CMT time. The resulting slowdown in decision making caused delays and frustration.

The six sequential steps of the NPDP, combined with the functional organizational structure, had the effect of pitting one functional department against another. The nature of the process was to pass the focus of activities from one group to the next, causing PMT members to make decisions that protected their own group. A minor decision, advantageous to one group, might present a large disadvantage to another, later in the process. The general feeling was that the functional vice presidents on the CMT were equally biased, leaving the president as the company tiebreaker. The PMT manager's role was a solution designed to address the issue of a PMT tiebreaker and provide each PMT with the ability to resolve problems quickly when necessary.

PMT managers are selected from the senior employees who are knowledgeable of Chipcom business and product priorities. They should also be aware of other issues, such as time-to-market, external agreements, cost issues, and product interdependencies.

Each PMT would have a PMT manager who would be a "virtual CEO" or the caretaker for the PMT, with the responsibility of selecting a PMT leader and work with the PMT to make sure it has the skills needed to bring a product to market. The PMT manager would be a resource for the PMT, using his or her experience to help the PMT, and when requested, would make critical product or business decisions. The manager would ensure that interdependencies are factored into the product priorities and would clear obstacles preventing the PMT from bringing the product to market. Most importantly, the PMT manager would act as a tiebreaker, in those instances when the PMT cannot resolve a problem internally.

NPDP Rewrite

The task team recognized that the existing NPDP document was in need of an update, given the changes that the PACT proposal recommended. A separate task team, including some members of the original PACT task team, formed to rewrite the document. This was done so that framework developed from the PACT discussions could be passed on to the new team, without going through all the original discussions. New members were included so that others might be brought into the process, developing further acceptance for concurrent engineering. Each of the functional department heads was asked to recommend members from their groups for the rewrite team. Some key members were requested to join the team so that there would be enough senior members on the task team to get essential input.

Time commitments for all the task team members were discussed prior to the first meeting. During the PACT discussions, the team met twice per week for 3–4 hours. This was a major commitment of time and resulted in people having to miss some of the meetings. The best method for resolving the time commitment problem was to concentrate the time commitment over 2–3 days, rather than spend 1–2 hours per day over several weeks. The rewrite team would meet off site for 2 full day sessions. The team had a technical writer assigned to it so that a professional would develop the final document. This would also free up the time of task team members to discuss the issues, rather than devoting one or two people to collaborating on recording notes and writing up the process.

The NPDP rewrite project was divided between Phase 1/Phase 2 and the Development Phase. A PACT member who also had a stake in the outcome of that part of the process was assigned to lead each of these groups. Team members divided up the work, with particular attention paid to assigning team members who represented each of the functional groups. A preliminary meeting was held with all the members of the task team, to present an agenda for the off-site meeting and to discuss the timetable for the rewrite.

When the task team gathered off site, they broke up into two groups to address the specifics of their section. The goal was to develop a first draft of the process, focusing on ideas rather than on the specific wording of the document, which would be filled in later with the help of the technical writer. Each area of the process was discussed, to identify the key elements of that activity, the responsible organization, and deliverables. Each group also discussed PMT membership and their responsibilities, because of its importance to the whole process.

The technical writer developed the first pass version of the document and circulated it to each team member. Several follow-on meetings were arranged for the purpose of going over each section of the process to ensure that the wording was accurate. Particular attention was given to the roles and responsibilities of the PMT members so that each member would have a clear statement of his or her contribution to the team. In addition to the specific responsibilities of each member that was particular to the member's functional skills, global responsibilities were defined. These included active participation, reviewing all product documents, reporting on project status, attending design reviews, updating the team on status and performance to schedule, and participating in PMT evaluation meetings.

The technical writer also brought the unexpected benefit of making the document very readable (rather than the cold, staccato approach of most technical procedures) through a cohesive presentation of material and the ability to highlight the most important aspects of the process. All recommendations for improvement or correction were collected, and a final draft was circulated to the CMT for approval.

TOOLS

CAD (computer-aided design), CAE (computer-aided engineering), and CAM (computer-aided manufacturing) tools can play a major role in implementing concurrent engineering, but must not be its focus. The organization should concentrate on the interdependencies of each group and how they relate to the new product process. Tools assist Engineering in doing it right the first time and in shifting the task from a serial to a parallel process.

The Electronic Data Automation (EDA) group, which supports the CAD and CAE tools, has assisted Manufacturing in developing methods to transfer CAD data directly to automatic placement equipment at the contract assembly house, resulting in the ability to reduce errors in programming and to accelerate the new product manufacturing ramp-up. Electronic Data Interchange (EDI) links can enhance relationships with suppliers by transferring purchase orders and payments electronically, accelerating the transaction and reducing their cost. These electronic relationships are examples of CAD/CAE/CAM integration and don't occur quickly because it presumes a greater level of trust beyond "arm's length" dealings between the corporation and its suppliers before either will allow electronic entry into each other's corporation.

The installation of workstations by groups outside of Engineering allows these groups (such as Advanced Manufacturing Engineer and Technical Support Engineer) to review design information while the design is in process. A concurrent design process requires access to design information, in real time, to effect change before committing ideas to paper or building prototypes. Manufacturability begins with the product specification and achieves the greatest benefit in reduced cost, improved quality, and reduced cycle time, when all the design members contribute early in the process. Incorporating manufacturability, serviceability, and testability later in the design process means that benefits diminish because of higher implementation costs and a lower probability that the PMT will risk modifying the design late in the schedule. Chipcom has outlined its manufacturability and testability rules in specifications controlled by Engineering Change Order (ECO).

Supplier relationships are a competitive advantage for corporations that can build the trust necessary to make them work effectively. LAN equipment has a high degree of complexity, with electronic design and testing of various communication protocols, double-sided surface mount boards with a mix of fine pitch, multi-chip modules, and all the processes needed to bring these products to market. A corporation needs to master a significant number of disciplines. Chipcom has approached its supplier relationships through a model similar to the "virtual enterprise strategy." [4] Chipcom has defined this strategy through the establishment of a Manufacturing Council. The Manufacturing Council consists of representatives from companies that Chipcom believes are essential to its operation. This might be the board assembly contractor, key suppliers, and strategic partners. Together they manage the entire value chain, resulting in the delivery of the product to the customer. These supporting companies have an expertise in their business that Chipcom utilizes. They make the investment to remain technologically current, and so Chipcom can take advantage of the latest technology. Chipcom is able to leverage the capabilities that each company has, such as people, relationships with other companies, and technical expertise. Additionally, these supporting businesses improve their businesses by learning to focus outwardly to satisfy customers, as well as inwardly to invest in improvements in equipment and processes. This type of relationship is established only with a small number of suppliers essential to Chipcom's business. A relationship of this nature cannot be maintained with a large base of suppliers, since the investment in time and resources is too high.

New EDA tools were incorporated into the existing environment, and employees were continually trained to upgrade their skills. Design managers were utilized, such as those developed by Mentor, that allow access to design documents while the designer is still working on the design. A design manager facilitates concurrent design methods by making work groups, libraries, and data management available to all users. It helps an organization break down barriers between groups through electronic connections. It addresses this through the distribution of design data over the network, administering and controlling part libraries, querying and developing reports in design data, and assembling and distributing correct documentation.

A "technology roadmap" was developed to identify the emerging technology and skills that the corporation would need and the means of acquiring them. This plan was jointly developed by Advanced Manufacturing Engineering and Engineering, who compiled this information through company visits, seminars, university liaisons, and so forth. No company can invest in all technologies discussed within the plan, but should use this tool as a method of forecasting future needs and to optimize the use of internal resources.

The technology roadmap is a look at the state of technology 1–3 years out. It seeks to shorten the product/process development cycle, when new technologies are introduced, by matching product needs with advanced technologies, tools, and processes. Figure 5-3 shows a possible outcome from one section in a technology roadmap focusing on substrates. The results are particular to each company, since the set of products, customer demands, competition, and pricing climate is different from company to company, resulting in different conclusions.

Figure 5-3 displays seven substrates considered for future PCB design and recommends that the hypothetical company invest in learning how to design with FR405. The reasons might be due to its higher glass transition temperature or to more even curing characteristics.

Utilize techniques of incorporating the customer's voice (an example is QFD). The foundation of this tool is designing the product to reflect

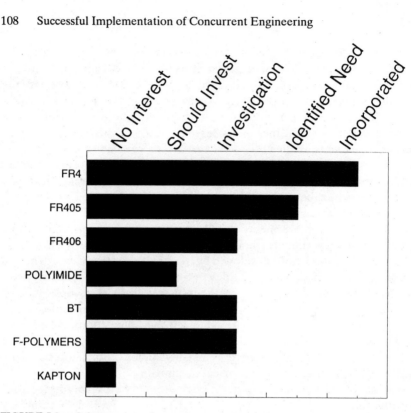

FIGURE 5-3. Substrates.

customers' desires and tastes so that Marketing, Design, Manufacturing, Technical Support, and Sales work together to formulate a product designed by customer response.

Software application tools, such as standard software packages, are ways of presenting information so that effort is not spent on converting information from one form into another. MIS can play a large role in establishing software standards and evaluating software packages. This also supports the availability of information within the organization. Common software and formats means that everyone can access information. Graphics packages, word processors, and project planning software are all essential for the smooth operation of the PMT.

Design for manufacturability (DFM) tools are an important factor in designing products that can be built cost effectively, with improved time-to-market and fewer changes after introduction to Manufacturing. The typical DFM tools had been a set of design guidelines that represented the ideal characteristics of a product in terms of testing, PCB layout, and component selection. New software products have been added to

Chipcom's definition of DFM through additions to the EDA tools that incorporate design rules into the CAD system and that add simulation capability. It is not possible to integrate all rules and guidelines into the CAD system, so some level of stand-alone tools is necessary. Chipcom's products are primarily PCB modules, so software that evaluates the PCB for density, auto insertability, manufacturing problems, and cost analysis has become a tool for Advanced Manufacturing Engineering, with potential roll-out to the hardware designers. The DFM software, such as that which Boothroyd-Dewhurst, Inc. markets for PCB, is a valuable tool for analyzing designs and providing an objective benchmark before laying out the design in CAD.

Design guidelines still remain the essence of our tools. These are constantly being review by Technology and Advanced Manufacturing engineers. Many are incorporated into the CAD system and focus on placing components and designing PCB. Others are available in written form and need to be understood by both the hardware and advance manufacturing engineers. Testability guidelines are a good example of those guidelines not incorporated into any automated system; therefore, their execution is dependent upon hardware designers and the Test Engineering group in Manufacturing to ensure that the guidelines are met through design reviews.

TRAINING

Training is a major focus within concurrent engineering at Chipcom and would include the PACT process, project management skills, QFD training, team building, leadership training, performance appraisal skills, and technology training.

An important aspect of training for Chipcom has been to train the existing design teams in the new NPDP. Since this training is very process specific, it was determined that this would be best delivered to the design teams by participants within the process. Members from the NPDP rewrite task team were chosen, since it was felt that they were the most familiar with the new process. A 1-day training class was set up as an introduction to the process. All participants were urged to read a copy of the NPDP document prior to attending the training. The session was videotaped, to be used for future training. On-going and more in-depth training would be delivered by HR in conjunction with participants in the process. A questionnaire was filled out by all participants to understand how best to improve future training, and follow-up meetings were planned for continued discussion around potential process improvement.

MEASUREMENTS

A concern of the task team was how to show that concurrent engineering was improving the new product process and not simply replacing one process with another. A second team was set up to address this issue and established five key metric categories. These measurements are not unique to concurrent engineering, but are metric categories that Chipcom would want to show consistent improvement in.

Customer Satisfaction

The goal of any business is satisfy the customer. The team viewed this as the primary metric category. Market share gains, warranty claims, customer complaints, on-time delivery, and percent reorder were selected as the specific measurement that would be set up.

People

Chipcom values are the basis for the method that the company uses to do business. They are detailed in a document that all employees receive when they join the corporation. The first section of the value statement specifically addresses our beliefs about the employees of the corporation: "We, the people of Chipcom, are proud of: our Company, our fellow employees, ourselves, the job we do for all our customers, and the reputation we earn by doing so..."[8] It makes sense that a second important metric category involves people and their performance. Several metrics that the team identified to track people's performances were percentage of on-time performance reviews, turnover rate, and percentage of work time devoted to training.

Product Delivery

This was the main area that Chipcom had started out trying to improve. Improving the product delivery cycle meant improving margins, improved return on investment, reduced run-rate development costs, less risk in forecasting the needs of the customer, and the ability to utilize the latest process and product technologies. Several metrics that the team identified for this category were the number of ECOs generated after a product was introduced, time-to-market as measured from phase 1 to production, design stability based on bill of material changes, and the number of part numbers.

Financial Health

A primary goal of the team was to show the potential return on investment for concurrent engineering. It was important to establish some metrics that addressed the performance of the company before and after concurrent engineering. Several metrics that the team identified for this category were return on gross assets, cash flow, asset turnover, and value added per employee.

CONCLUSION

Concurrent engineering is a process. It works for Chipcom because the company identified for itself the internal relationships that allow it to produce the best products for its customers at the right time. Chipcom then practiced those relationships, to turn them into "the way the company does business" and not just another program. It also worked because concurrent engineering focuses on the team approach. Chipcom already had the PMT established as the means by which it developed new products. Concurrent engineering built on the culture that was present, redefined relationships, and clearly delineated responsibilities. Each PMT member had to accept accountability for his or her own decisions and to understand how his or her responsibilities supported the team. Each member had to understand that the commitments made to the team needed to be completed on time or else delays had to be communicated accurately to the team. The team members had to respect each other's skill and work together towards a common goal, negotiating solutions acceptable to the team.

EPILOGUE

The initial effect of implementing these changes was a reasonable level of confusion as the teams changed to the new process. In particular, most teams were unclear as to which documents needed to be generated by the team to get products approved by CMT. Since a cross-functional team was used in developing the new NPDP, there was generally one member on each PMT to lead the team through these changes. The viewpoint by most people that participated in the new process was that the new process simply represented changes that were already underway, without being stated formally. The changes made sense for Chipcom, at this time because of its rapid growth. There was a feeling of urgency to develop new processes that would support the accelerated growth and dramatically reduce the time-to-market. The company also started a PACT PMT to focus on issues needing on-going attention to support concurrent engineering and to review recommended changes to the NPDP. Chipcom is over a year into concurrent engineering. The changes that Chipcom has made are working and have helped to focus the company on clearly defining each product before beginning the design. It is also intent on seeking the input from all functional groups, customers, and suppliers. The company also realizes that the most recent changes to the NPDP is but one step in "n" revisions of the process. Change has to be continuous to be effective.

DISCLAIMER

Any opinions or views expressed in this chapter are the opinions and viewpoints of the author and do not represent the opinions or viewpoints of Chipcom Corporation.

References
1. Chipcom 1991 Annual Report.
2. Chipcom Market Communication, Chipcom Corporate Overview, 1991.
3. Concurrent Engineering Proposal, "Partnership for Achieving Champion Teamwork," Concurrent Engineering Task Team, December 1991.
4. Flaig, L. Scott. 1992. The virtual enterprise: your new model for success. *Electronic Business*. March 30, 1992, pp. 153–155.
5. Herner, Alan E. 1991. *Wright Laboratory Initiatives in Integrated Product Development*, Proceedings, Nepcon West, p. 935.
6. New Product Development Process, Chipcom Corporation, Rev C, 1991.
7. Whitney, Daniel E. 1988 Manufacturing by design. *Harvard Business Review* 7/8, p. 83.
8. Chipcom Corporation's Values.

6
Concurrent Engineering: Sun Microsystems

Christopher Natale
Sun Microsystems Incorporated

PURPOSE

The purpose of this chapter is to explain the importance of concurrent engineering and its impact on product development success. Several examples of how Sun Microsystems practices concurrent engineering are also presented.

INTRODUCTION

Sun Microsystems Computer Corporation is the world's leading supplier of open client-server computing solutions. The headquarters are located in Mountain View, California, and Sun Microsystems has manufacturing plants in Milpitas, California; Linlithgow, Scotland; and Westford, Massachusetts. The company was founded in 1982 and has grown into a multi-billion dollar corporation in 10 years.

Despite difficult economic times, Sun Microsystems has continued along a successful path. Revenues and sales have grown steadily. Why is Sun Microsystems doing so well? This chapter will answer this question as well as illustrate the effects of applying various concurrent engineering principles. This chapter focuses on the operations of the East Coast division of Sun Microsystems. The East Coast division is comprised of an engineering facility in Billerica, Massachusetts, and a manufacturing facility in Westford, Massachusetts.

CONCURRENT ENGINEERING APPLICATION IN THE NEW PRODUCT DEVELOPMENT

There are several major reasons why Sun Microsystems is doing so well. We deliver:

- High-quality products that are competitively priced.
- Products that utilize leading technologies.
- New products to the marketplace in a timely fashion.

The application of concurrent engineering principles helps us to achieve these critical success factors. Concurrent engineering, also called simultaneous or parallel engineering, is the process in which many functions are being performed simultaneously. This process cuts new product introduction (NPI) developmental time exponentially. Concurrent engineering is a cross-functional interdisciplinary activity that begins at the prenatal stages of design and continues through production and product end-of-life. Unlike the traditional "serial" new product development (where one action had to be completed before another begins), concurrent engineering allows multiple activities to begin before prerequisite activities are even started.

Figure 6-1 illustrates the differences between concurrent engineering and sequential engineering.

The main reasons why concurrent engineering cuts developmental time are:

- Multiple actions are being performed simultaneously.
- If one item in the new product development process slips from the schedule, it will not hinder subsequent activities.

SUN MICROSYSTEMS NEW PRODUCT INTRODUCTION MODEL

The proper organizational model must be established to support a successful concurrent engineering effort. Concurrent engineering requires an abundance of cross-functional activity by all NPI team members. Upper management must be fully committed to supporting this philosophy and culture.

Figure 6-2 illustrates the communication flow and interactions of the various organizations within the NPI process.

Sequential Engineering

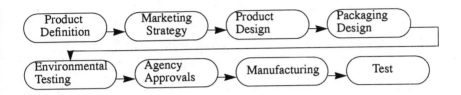

Concurrent Engineering

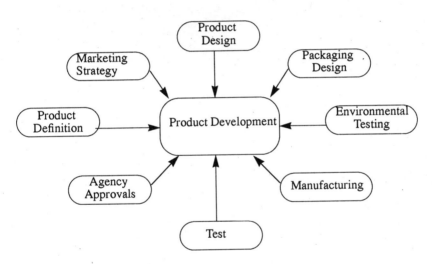

FIGURE 6-1. (a) Sequential engineering; (b) concurrent engineering.

SUN MICROSYSTEMS BUSINESS TEAM AND PRODUCT TEAM MODEL

The communication interaction that is shown in Figure 6-2 is achieved through the use of a business team and several functional teams.

The business team is established in the conceptual stages of a new product and prior to the "official product approval." This business team is comprised of the following people:

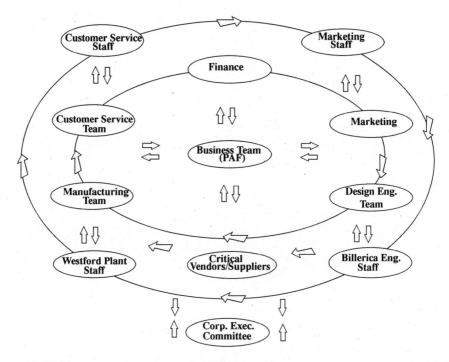

FIGURE 6-2. NPI program information flow.

- Team leader, who is usually a director or vice president from Design or Marketing.
- Design manager.
- Manufacturing program manager.
- Customer service manager.
- Marketing manager.
- Finance manager.

The role of this business team is to continue with the investigation of the proposed product and to create a formal product plan called a Product Approval Form (PAF), which is submitted to an executive-level Product Strategy Committee.

The PAF is a very detailed document that consists of:

- Executive overview.
- Marketing plan.
- Product overview.
- Competitive overview.
- Product development overview.

- Operations overview.
- Customer service overview.
- Business plan and strategic assessment.

The PAF defines why the product should be developed and how it will be marketed, designed, and manufactured. This document will also detail product developmental costs, the target first customer ship (FCS) date, and the approximate sale price.

Once the Product Strategy Committee approves the new product, an NPI team is formed. In many instances, the core NPI team is established prior to the "official" product approval. This allows the team to get postured for the new product and to provide inputs into the PAF.

The business team and functional teams are shown in Figure 6-3. The functional teams are the Engineering Qualification Team, Operations Team, Marketing Team, and Customer Service Team. The business team acts as a "steering committee" for these functional teams. The members of the business team become the "program managers" on their respective "functional" teams. Figure 6-3 illustrates the interrelationships of these teams.

ROLE OF THE NPI TEAM MEMBERS IN THE CONCURRENT ENGINEERING CAPACITY

In the product team environment, the team members have various roles. This section outlines the team member responsibilities in the new product development process. The operations team forum provides a communicative environment to set team objectives. The structure of the operations team is illustrated in Figure 6-4.

Following is an explanation of the team members' roles:

- Role of the NPI program manager—This person has overall responsibility for coordinating all manufacturing activities in the new product development cycle. This individual also participates on the cross-functional business team that defines product and schedule.
- Role of the marketing representative—The marketing representative determines the marketing strategy for the new product, and defines the market for which the product should be targeted. The marketing representative performs a market trend analysis and studies technology trends.
- Role of the manufacturing engineer—The manufacturing engineer's primary function in the new product development process is to ensure that the new product meets high-volume manufacturing requirements. Products must be easy to assemble. The manufacturing engineer also defines and establishes the optimum manufacturing process.

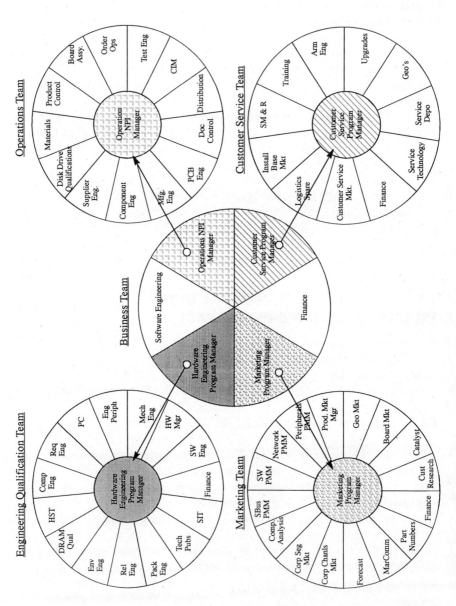

FIGURE 6-3. Business team communication model to functional product teams.

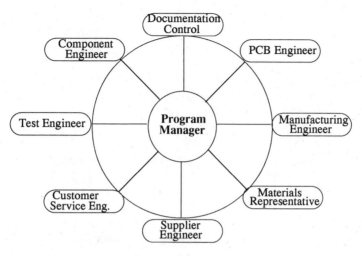

FIGURE 6-4. NPI operations product team.

- Role of the supplier engineer—The supplier engineer's role in the new product introduction process is to work closely with the supplier to ensure that they have adequate facilities and processes to produce a high-quality product. Activities include performing supplier surveys, certifying the assembly and test processes, and defining test requirements on Sun platforms at the end of their process.
- Role of the design engineer—The design engineer works closely with mechanical and manufacturing engineering to develop a product that meets the product requirements as defined in the product approval package. Design engineers interface closely with industrial design and environmental engineering to ensure that the product meets its other critical requirements.
- Role of the customer service representative—The customer service engineer works to insure that enclosures are designed to be customer maintainable and to insure that there are on-board and stand-alone tests capable of isolating failures. The customer service engineer also provides information about new technologies that could impact service delivery.
- Role of the test engineer—The test engineer develops and institutes the test processes for the new products. This person interfaces with design engineering to ensure product testability.
- Role of the PCB engineer—The PCB engineer works closely with the board design team to ensure that design for manufacturing standards are integrated into the board design. This engineer also works with manufacturing to set up an appropriate manufacturing process.

- Role of the component engineer—The component engineer works very closely with design engineering to define the appropriate components to be used. The component engineer ensures that the suppliers' component specifications meet our requirements.
- Role of materials acquisitions representative—The materials acquisitions representative works closely with design engineering to determine the suppliers. The representative works with the suppliers to obtain competitively priced parts, lead times, and payment terms. They also facilitate the procurement of parts required for prototype and production builds.
- Role of the production control coordinator—The production control coordinator is responsible for scheduling the prototype and production builds of new products. This person also ensures that the material is available for these builds.
- Role of the production supervisor—The production supervisor ensures that the manufacturing and test processes are established and documented clearly and concisely.
- Role of the documentation project coordinator—The documentation project coordinator's main function is to help define bill of material structure and to manage engineering change activity and pre-engineering released products. He or she also manages prototype material requirements and facilitates engineering release ECO.

PRODUCT PIPELINE

The product development model shown in Figure 6-5, called the product pipeline, illustrates the new product development process, from product conception through product end-of-life.

PRODUCT DEVELOPMENT CYCLE

Figure 6-6 illustrates the product development cycle.

CONCURRENT ENGINEERING EXAMPLE 1: COMPETITIVE ANALYSIS

One of the aspects of concurrent engineering that we apply at Sun is competitive analysis. The Advanced Manufacturing Technologies group purchases various "best-in-class" competitors' products. A competitive analysis team is formed of representatives from the following groups:

- Customer Service.
- Supplier Engineering.

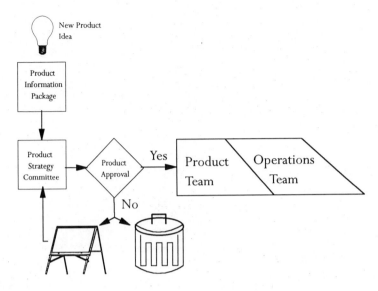

FIGURE 6-5. Product pipeline.

- Test Engineering.
- Commodity Management.
- Manufacturing Engineering.
- Product Marketing.

The competitive analysis team performs a thorough product analysis and disassembly of these products. The analysis includes an in-depth study in the following areas:

- Cost.
- Service.
- Test.
- Delivery.
- Time-to-market.
- Warranty.
- Quality.
- Design for manufacturability.

The team representatives compare our products to our competitors' products. The favorable and unfavorable product attributes are examined. This information is then combined, to make a complete competitive analysis report. The report, which details information down to the component level,

122   Successful Implementation of Concurrent Engineering

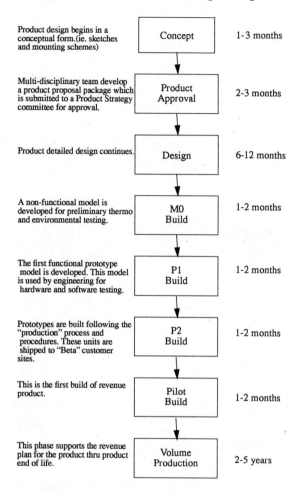

FIGURE 6-6.   Product development cycle.

is distributed throughout various organizations in the company. The report contains:

- An executive summary.
- Conclusions that summarize the bottom line results.
- Comments on various favorable design characteristics.
- Recommendations to be considered on our next new product.

One area of the competitive analysis that is particularly useful is the design for manufacturability (DFM) study. The manufacturing engineer disassembles the unit and performs DFM analysis to the component level. By applying the Boothroyd Dewhurst analysis method of DFM, a quantitative design efficiency is obtained for each product.

When the manufacturing engineer performs the DFM analysis, he or she discovers various ways to modify our existing products, to improve product manufacturability. The manufacturing engineer performs "what if" and feasibility studies to determine benefits of subassembly redesigns.

CONCURRENT ENGINEERING EXAMPLE 2: DFM IMPLEMENTATION INTO NEW PRODUCT

A very critical aspect of new product development is ensuring that the new product is designed for manufacturability and assembly. A very effective method of achieving this objective is for manufacturing to get involved in the product design process. At Sun, the manufacturing engineers begin meeting with the product design engineers very early in the design stage. Detailed discussions begin in the conceptual stage. During this time, manufacturing communicates their requirements and works with the design engineers to come up with design solutions that meet both product requirements and producibility requirements. Figure 6-7 illustrates this manufacturing/design engineering communication activity.

Since 80% of the product cost is locked in at design, the benefits of this early involvement is substantial. By meeting in this early design stage, manufacturing can get their assembly requirements built directly into the product design. Hence, when the product is released, there is no need for product modifications, redesigns, or cost reductions—all these favorable design changes are incorporated in the design. This design/manufacturing relationship has proven to be very successful within Sun Microsystems. With every new product, development time is shorter, products are better, and the design/manufacturing link gets stronger and more effective. Figure 6-8 illustrates our current design/manufacturing link. This figure also represents our continual goal and efforts to tighten this relationship.

CONCURRENT ENGINEERING EXAMPLE 3: PREVENTATIVE MEASURES TO ENSURE TIMELY PRODUCT AVAILABILITY

One of the basic underlying principles of concurrent engineering is to meet critical program target dates. This case study illustrates an example of how Sun ensured product delivery by the product release date.

FIGURE 6-7. Design engineers and manufacturing engineers meet in the conceptual stages of the initial design activity to ensure a highly manufacturable and producible product.

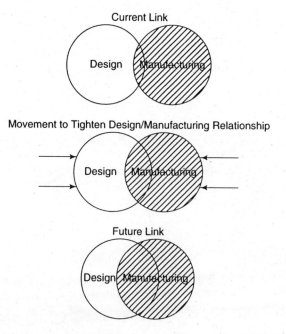

FIGURE 6-8. Design/manufacturing link.

The Sun Microsystem's "Sunergy" CPU product had a customer announcement date of November 10, 1992. This was a critical date that had direct revenue impact and could not slip. All mechanical and functional aspects of the product were well within the program plan. There was, however, one area that had potential exposure. This area was product agency labeling.

The product has a multitude of agency approvals, including UL, CSA, TUV, VCCI II, FCC Class B, and Sparc Compliant 2.0. From a DFM perspective, these agency approval logos are combined together to create one product label. The problem was that we could not obtain a firm commit date on when we would receive the official approvals from two of the agencies: UL Class B and the Sparc Compliant 2.0. Labels must be put on our units, yet we cannot apply "agency approved" labels unless the product is indeed agency approved.

This labeling issue and the inability to obtain a firm approval date posed quite a serious problem—a problem so significant that it could jeopardize the first customer ship date and paralyze product manufacture, which would result in substantial loss of revenue and market share. Ten weeks before the product announce date, and still not being confident about the agency approval coming in on time, the materials representative, the manufacturing engineer, and the product design engineer purchased an insurance policy to eliminate any fears relating to the agency approval dates.

What kind of insurance policy was this? We created our own insurance policy. We developed three different sets of label artwork and purchased three different product labels with the different combinations of agency approvals. This ensured that regardless of which and when agency approvals were granted and when, we would be shipping our new product to customers by the November 10, 1992 product announce date.

This exercise cost in the vicinity of $2,000, but it was money well spent. The solution in this case study was simple yet extremely effective. In the new product development model, there are key dates that can't be missed. It is the responsibility of the team members to highlight potential problem areas and communicate these to the team. The team must address and resolve critical issues to ensure that product meets important product development dates.

7

From Specification to Beta Site in Three Months: Concurrent Engineering at Mercury

Allen Zubatkin
Mercury Computer Systems, Inc.

Mercury Computer Systems, Inc., a producer of embedded high-performance, floating-point computing systems, typifies progressive company agility, with responsive local subcontractors and procedural innovation conducive to product design and development. The chapter headline refers to the MC860VM, a single processor that utilizes the Intel 40MHZ i860 microprocessor. It is a 16-Mbyte memory subsystem, with a local high-speed direct I/O bus, a plug in interface for an additional add-on card, and a VMEbus interface. It delivers up to 80 megaflops of single precision or up to 60 megaflops of double-precision floating-point performance. This product development chronology was outlined in an article by Dozier and Yacovitch in *Surface Mount Technology* magazine, April 1990.

The product is complex both in design and manufacturing technology. It is comprised of an 8-layer printed wiring board (PWB), including 111 surface mount devices (SMDs) requiring 1,238 pads and over 3,000 through holes supporting insertion mount components (IMCs) and via interconnections. Moderate fine line (0.007-inch) trace/track technology was used, with double track in the through hole and single track in the surface mount regions. Manufacturing performance results were stellar, as the first lot of 20 had 100% yield in final test.

The company is currently working on products that are two generations ahead in performance to the MC860VM, and is transitioning into fully integrated systems, with hardware and software qualifications and a solution approach to customer needs.

THE MERCURY ADVANTAGE

The enabling factors for success are typical of the advantages of concurrent engineering, magnified by the company's operating philosophy. They include the following:

- A flattened, lean organization, with a minimum of layers to work through, at all functions. This philosophy keeps the overhead low, with advantages realized in cost, time, and faster communication.
- Lots of communications at all levels. There is a minimum of formal meetings and many working sessions, concentrating on problem solving, strategy, and trade-offs necessary for achieving optimized project plans: resources, schedules, and scope of the projects.
- The product development process, including the product definition (specifications) and the new product introduction process, is well established and practiced by the company. It is a flexible process, depending upon part availability, cost, margins, performance, and the trade-off of these factors.

Manufacturability, testability, reliability, and time-to-market are critical issues that were integrated into the product design process. Towards that end, a flexible manufacturing process was developed that maximized the quality of all products, including nonstandard and custom solutions. Manufacturing employs a strategy of mutually beneficial long-term relationships with a small number of suppliers and subcontractors, to allow for quick turnaround of prototypes and production runs.

A special Advanced Manufacturing Engineering group (AME) was entrusted to set down manufacturability guidelines for designing product, qualifying and monitoring subcontractors, and coordinating these functions with other internal company activities. In this capacity, AME acts as a watchdog, ensuring that practical limits imposed by current production technology are observed within the physical design of new products. In addition, AME is continually researching and updating the design and manufacturing knowledge base for new ideas, technical advances, and procedural innovations that can speed up production, reduce cost, and enhance product performance while maintaining the high quality expected in the product.

At the same time, prototype/quick turn is not an excuse for multiple redesigns. The product planning criteria allows for only one redesign beyond the prototype phase. In many cases, the prototype PWBs have become the final product.

THE PRODUCT CREATION PROCESS

Original research and development is performed by the Advanced Product Planning group (APP), which is a long-range technical group chartered with new concepts and technologies. Once a product opportunity is identified, the marketing department takes over through the Product Management (PM) function. PM assumes the coordination between all the other departments and makes sure that the timelines and critical paths are all negotiated and defined. Each contributing department will fill out its portions of the project plan and fold it into the overall product rollout schedules, as labeled.

The traditional line between engineering and manufacturing in the development cycle is nonexistent and was planned in this manner. Engineering is very proficient at developing architectures, circuit designs, and solutions for technical problems. They are inherently less interested in performing tasks that require a high level of control and repeatability, which happen to be the very elements that make a manufacturing organization successful. It is best to dovetail both of those capabilities and strengths together so that the combined result is greater than any individual effort.

MERCURY'S MANUFACTURING PHILOSOPHY

The manufacturing strategy at Mercury is based on the utilization of subcontractors and suppliers whenever possible, in order to release capital resources for research and development. In addition, the electronic industry technology moves so quickly that the type of equipment needed to manufacture today's products cannot keep up with the technology of the next generation. A positive *return on investment* (ROI), based on the company's production volumes, would not likely ever be realized.

The direct and tight coupling between new product development and manufacturing resulted from a discussion that occurred early in the company's history. Manufacturing suggested that the purchasing department buy all prototype materials and components. Both parties benefitted: Development received good service and the benefit of purchasing experience with marginal suppliers, and manufacturing got insight as to what parts will be purchased for the next set of products. Since then, there has been a constant evolution and improvement of the new product introduction (NPI) process. Currently, manufacturing continues to do all the purchasing for prototypes in research and development, as well as perform/ support all mechanical design.

Throughout all new product development projects, manufacturing personnel work along with the designers, who specify integrated circuits and architecture and decide with the CAD operators where the parts should go and how best to manage air flow, mechanicals, and cooling mechanisms. Issues and trade-offs in such topics as packaging, manufacturability, testability, time-to-market, cost of introducing new products, risks, and how to ramp to full production and producible quantities as quickly as possible are discussed and agreed upon.

This NPI process methodology can be presented from the following functional perspectives (see Figures 7-1 to 7-4):

1. Material acquisition—Including material control, logistics, and acquisition.
2. Manufacturing engineering—Including how the product is designed, from handoff through full production.
3. Test and diagnostics—How will the product be tested, what is the confidence level, and how will testability, quality, and reliability be imparted to the design.

Material Acquisition

This task is initiated once engineering feels that they have assembled a preliminary bill of materials (BOM). The BOM is sent to manufacturing in two forms: One specifies a source of supply and the other specifies a generic manufacturer. The BOM is reviewed first with the specified suppliers. Are

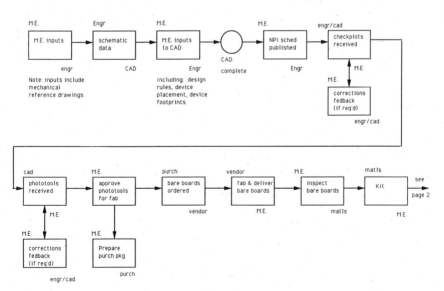

FIGURE 7-1. New product introduction process—manufacturing engineering.

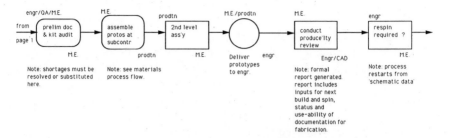

FIGURE 7-2. New product introduction process—manufacturing engineering.

they approved? Does the company have any history of difficulties, either of a technical or of a business nature, with any of the supply sources? If such issues are raised, manufacturing will get back to engineering and suggest that they find another source, if possible.

For components that do not have sources of supply, suppliers are identified for parts that are technically compatible for the application and engineering sign-off is obtained, including part numbers. Each specific board is assigned a project number. There might be more than a dozen products being designed simultaneously, and costs are estimated for direct materials, labor, and engineering so that ongoing development costs are measured.

The next phase is the acquisition of the material from the preliminary BOM, after enough suppliers and components are identified. The timing depends on the scope and the status of the project. Shortages and availability problems will always exist, and purchasing will have to either expedite or substitute parts supplied from engineering. At this point, the manufacturing

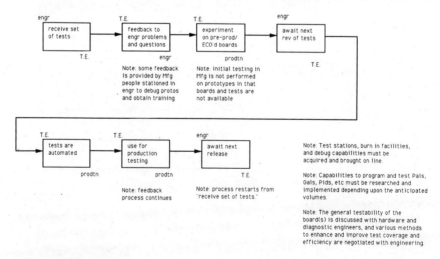

FIGURE 7-3. New product introduction process—test/diagnostics.

From Specification to Beta Site in Three Months   131

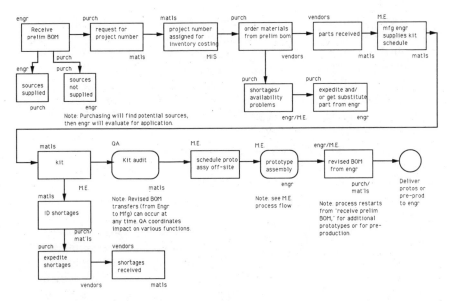

FIGURE 7-4. New product introduction process—material acquisition.

and engineering functions determine the kitting schedule, which is the number of prototype boards that are going to be built. The materials organization will assemble kits for those specific quantities.

During the kitting process, shortages are identified again and purchasing has another phase of expediting and/or finding substitutes. It is an ongoing process that is reviewed at two or three critical stages of the prototype build. After a kit is completed, it is audited by quality assurance to make sure that it is up to the current revision of the preliminary BOM.

At this phase, product performance and/or product characteristics are typically changing, giving the impression of trying to hit a moving target. Prototype assembly is scheduled at an off-site contracting facility, where many types and levels of technology are being scheduled at the same time. Only a portion of each prototype kit is assembled so that, when additional prototypes or product changes occur, they can be built expeditiously, with only minor modifications of the remainder of each kit.

Manufacturing Engineering

Manufacturing engineering activity starts when there is enough product concepts, architecture, and sense of the package size from the marketing group. Some initial working drawings are created, either by design engineering or manufacturing, that would record the size and location of major components, as well as any special or unique mechanical features. Initial BOMs are messaged, while the drawings are continuously updated. At some point, a stake is driven in the ground and the product concept is frozen: *"That's it! We are ready to go out and have this particular configuration designed."*

An outside subcontractor is used for all formal CAD layout and post-processing. It allows for many designs to be performed simultaneously. Close working relationships have been established with several of the local design houses, with good success. Typically, the design engineers will have to spend some time with the CAD layout engineers before manufacturing gets involved, to make sure that all of the components, documentation, and net lists are accurate. At the same time, manufacturing will start transferring inputs, such as design rules, device placement, footprints, and mechanical dimensions, and add that to the overall design within the first few days. This is an ongoing process between hardware design engineering, manufacturing engineering, and the CAD house to complete the design. Depending on the product, there may be a case where the design engineer will need to have more interaction with the CAD house because of routing and other schematic issues. In other instances, manufacturing engineering will have more interaction because of the mechanical layout and manufacturability issues. There are no hard or fast rules. Everyone understands what their role is, and the project goals are more important than those of the individual. Keep in mind that effective communication is the key to this overall success.

When the CAD operation is complete for the prototype, the schedule is updated, to ensure that everyone in the project understands the hardware issues from the software and documentation perspective. Check plots are inspected by the manufacturing engineers, to make sure that the routing does not violate any of the design and manufacturing rules and that it is free of errors. If there are corrections to be made, they are fed back to the CAD house, and new check plots are cut and reviewed before final approval.

Once this milestone has been accomplished, hard prototype tools are produced and, when received, are reviewed by manufacturing and design engineering to make sure that nobody has missed anything and that they will produce high-quality PWBs. When the prototype tools have been approved, a package is prepared, including manufacturing fabrication guidelines, the photo tools themselves, and associated magnetic media information. They are delivered to a previously approved, quick-turnaround PWB fabrication house. Depending on the product, there may be some interaction between the vendor and manufacturing engineering for clarification. In some cases, the tools will change, based on the PWB fabrication vendor's inputs and comments. In other instances, the PWB vendor, as well as the CAD subcontractor and the assembly subcontractor, will contribute to the design before their part of the project is required.

Based on the price/performance required, six to ten prototype PWBs are built and received in about a week's time. They are inspected in detail by quality assurance and manufacturing engineering and the production rework technicians, since they most likely will be working on these boards in the future. At this point, the material kit will have been completed as previously discussed, and the kit is audited by four sets of groups: the materials group, quality assurance, design engineering, and manufacturing engineering. All parties want to make sure that they are building the right thing at the right time and the right quantity, with nothing falling through the cracks. If there are any shortages, they should have been discovered and resolved, or will be shortly, to support the build.

Assembly subcontractors are used to build the prototype PWBs. Again, these suppliers have been carefully selected so that Mercury has the option to build boards at the subcontractors facility very rapidly, with only the minimum documentation necessary. This arrangement also allows the subcontractor to move up the learning curve quickly, as well as to be integrated into the Mercury process. One example of this is the integration between the CAD house and the contractor assembly process: Components are laid out by the outside CAD house so that they can be automatically placed by the subcontractor's automated machinery as much as possible. In addition, the CAD house has most of that manufacturing data on line and can have the data transferred electronically, if required, to the other subcontractor to produce the production tooling (e.g., solderpaste stencils for the assembly contractor who does the component placement): Typically, the assembly can be achieved in less than 2 days. In some instances, an arrangement has been made to allow Mercury manufacturing engineering to personally assemble the prototype boards, using the subcontractor's equipment and being assisted by his or her employees.

After delivery of the assembled PWBs, a complete visual and manufacturability audit will be performed by the manufacturing engineers, and rework technicians and the production team will audit the boards. The audit will use the same guidelines as the production PWBs: *"If you can build it right the first time, you can build it right all the time."* Manufacturing engineering will complete a part producibility review. This is a formal report that includes inputs for the next revision of the PWB, if one is required, and includes the status and usability of the current documentation for fabrication and assembly.

The design engineering department will then spend several days or weeks validating the prototype PWBs to the design specifications. During that process, production technicians are transferred to engineering, to assist in that effort and to begin the learning process for full production. If a redesign or a revision to the design is required, the process starts over from the very beginning and the same methodology for control and update of the schedule and timelines is adjusted and communicated to various organizations throughout the company.

Test and Diagnostics

The test and diagnostic process begins at the very early stages, where test engineering, diagnostic engineering, and hardware design engineering all discuss the kinds of trade-offs in the design and what each group would like to see accomplished. Depending on the trade-offs of resources, schedules, and desired performance level, a task list is generated. The diagnostic engineers are responsible for writing tests to exercise the product. The hardware design engineers assist them in achieving that task. The test engineers take the input of both groups, investigate the design, and report back on the test coverage, test efficiency, and what steps are needed to improve both. All of these events are scheduled to be completed as early as possible in the design cycle.

These steps are repeated during the second prototype and preproduction runs. It is an iterative process that is aimed at continually improving test coverage and efficiency. At some point, the tests are automated and loaded into a package as a set of test tools. The test operators can use these tools efficiently in both test and troubleshooting of the products. The most important aspect is to try to load the process up front as much as possible, to minimize any downstream problems.

SUPPLIER PARTNERSHIPS

Supplier partnerships are an important part of the overall Mercury competitive strategy. The development of a coordinated plan of action to locate, evaluate, and maintain supplier partnerships is of critical importance to the success of the company and shorter time to market. Several steps have been identified to ensure success in supplier partnerships and were introduced in the article by George J. Koslosky, "A Contractor Selection Checklist," in the August 1992 issue of *Surface Mount Technology* magazine, Vol. 6, No. 8, pp. 62–63.

These steps are:

1. Utilize a supplier evaluation team.
2. Use a defined process for supplier evaluation.
3. Help a Supplier exceed expectations.

Utilize a Supplier Evaluation Team

A team is formed, with representation from materials, manufacturing, engineering, and quality. The technical and materials expertise is necessary to assess and form both a technological and a business alliance. The quality representation is important, with the growing emphasis on statistical process control (SPC), formal inspection criteria, and other upcoming regulations, such as ISO 9002. The quality organization is best qualified to communicate the company's quality requirements and expectations to prospective suppliers.

The purchasing department, in coordination with the supplier evaluation team, should determine a relevant supplier rating, which is a key factor in maintaining and improving supplier relationships.

Use a Defined Process for Supplier Evaluation

A process approach was developed for the supplier evaluation team. The process begins with the determination of Mercury's own needs and expectations. These include examining the processes, technology, materials, volumes, and costs required to develop new products successfully. Once this process has been consummated, the evaluation criteria for each potential supplier is determined, with each member of the team completing the portion of the evaluation related to his or her own expertise.

A list of potential suppliers is then accumulated, with information coming from various sources: sales representatives, trade journals, purchasing, trade shows, directories, and recommendations from associates, both from within and outside the company. Proximity of the supplier plays an important role, since quick turnaround and personal communications are important.

Site evaluations and inspection by the team then follow. The purpose of such activities is to determine criteria for judging potential suppliers in:

1. Their technological competence.
2. Their management fit into the company's procurement strategy.
3. Their financial stability.

Intangibles, such as the supplier equipment inventory, their processes and procedures, and their adherence to world class manufacturing and quality standards, can be quickly ascertained. Evidence of SPC, just in time (JIT), and other quality methodologies should be readily available. Handling of *engineering change orders,* segregated customer storage areas, and coordination of shortages are critical issues to be examined at site visits.

A *request for quotation* is then issued to potential supplier candidates. All necessary documentation is supplied, and, if possible, applicable issues for manufacturability review are included. Realistic quantities, delivery times, and tooling charges are itemized, and premium conditions are determined.

Supplier selection is determined after careful analysis of the issues of technical competence, quality, pricing, and delivery. A trial job is contracted, and the supplier team purchasing representative assumes the role of the project manager during this phase of supplier evaluation. Good communication is the key to success during this initial trial. The purchasing representative should also assume the role of the communication "traffic cop" between the various experts working for both the company and its supplier, even for technical issues. In this manner, the severity of problems and their resolution can be examined after the job is completed.

The team holds a meeting with the supplier after the completion of the trial job to discuss the initial results, such as which elements went smoothly and which need improvement. The level of commitment and expertise shown by the supplier, as well as their quality level and their communication channels, are examined and discussed for future enhancements.

Help the Supplier Exceed Expectations

After the supplier has successfully completed the trial job, the business relationship with the supplier can be maximized by helping them exceed expectations. Some of the "help issues" can be in the following areas:

- Material and documentation are prepared and maintained correctly. Examples include maintaining and delivering to the supplier the latest information regarding engineering changes and bill of materials changes.
- Delivery dates and pricing are expected to be reasonable and correct. The volume and frequency of order variability should be kept to a minimum.
- Feedback is provided to the supplier on their performance, by communicating inspection results in a timely, accurate, and useful manner. This allows the supplier to correct process variations as early and cost effectively as possible.
- Official and regular communication links are established. These links are minimized in number to a few key individuals, to reduce misunderstanding, while unofficial links are eliminated. Regular feedback meetings are held with clear agendas. A proactive posture is important to avoid problems before they occur.
- Expectations are set jointly, especially with regard to the quality, workmanship, and inspection standards. Official standards, such as IPC or MIL specifications, are referenced clearly.

These procedures have helped Mercury Computer Systems establish a strategy of mutually beneficial long-term relationships with their suppliers, for quick introduction of high-performance, superior-quality, and low-cost products.

8
Design for Manufacturability at Northern Telecom

John Dransfield
Northern Telecom Canada Limited

Northern Telecom designs and manufactures telecommunications equipment in North America. Its headquarters is located in Mississauga, Ontario, Canada. Northern Telecom is part of a tri-corporate structure composed of Bell Canada Enterprises, Bell Northern Research, and Northern Telecom. Northern Telecom has worldwide manufacturing operations, with revenues of more than $5.5 billion, 60–70% of which come from products introduced since 1986. Northern Telecom annual new product growth rate is estimated at 15%. Effective new product introduction (NPI) capability is necessary to support its growth.

In earlier years, product life typically spanned many years. This gave the company the luxury of time to enhance designs, develop new manufacturing processes, and improve the cost and performance of new products. Recently, the company is faced with much shorter product development time to meet the windows of market opportunity, and must perform these tasks on a *do-it-right-the-first-time* philosophy.

The accelerating rate of technological change in printed circuit board (PCB) design and manufacture has prompted Northern Telecom to recognize an urgent need to provide the designers and engineers with specific design requirements and manufacturing constraints. This information helps the designer select the right design for the targeted manufacturing location, in order to meet the market opportunity at the lowest cost and highest quality.

Typically, 80% of the cost of a product is fixed at the first design stage. From the start, the designer has a direct impact on product cost, quality, performance, reliability, introduction intervals, manufacturing process yields, future maintenance, and even the end of life and disassembly of the product.

Northern Telecom has developed a successful strategy for new product design and introduction into manufacturing:

1. The *Gate process*—A formal, staged, and phase reviewed new product development process.
2. The *New product module*—A project management group dedicated to the introduction of new products.
3. The *Manufacturability assessment*—A tool to rate and feed back to the designers a measure of the manufacturability of new products.

THE GATE PROCESS

Northern Telecom has developed a new product development gating process. This process coordinates the activities for the successful transfer of new products from the design community through engineering, marketing verification, and manufacturing environments to the customer (Figure 8-1). This is accomplished by using a framework of project deliverables partitioned into timely milestones. These requirements facilitate the planning and control of resources for the timely introduction of new products. The objective of the gate process is to couple the development program to the needs of the marketplace and the business. A "gate" is a formal meeting to assess the

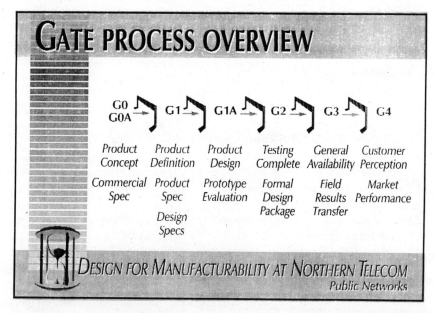

FIGURE 8-1. Gate process overview.

status of a product at a strategic milestone. Formal gate reviews manage the risk inherent in any new product development.

A team approach is used to coordinate the many activities of the different groups involved with the introduction of new designs. All deliverables from each gate must be met before the next phase of the project can proceed. Sometimes this process is known as phase review project management, where the project is partitioned into easily identifiable segments, and management sign-off is required to make sure all parties are in tune throughout the project.

The assessment of the new design is a progressive, interactive process, commencing with gate 1 and using on-line files in concert with design guidelines built into a corporate CAD tool. The requirements of the NPI process reflect the scope of group tasks, which can be met only through effective teamwork.

A minimum of five gate reviews are convened by the project director for most projects:

Gate Review 0—At a very early stage, when a basic idea is passed to Marketing for evaluation.
Gate Review 1—When major responsibility of the new product is transferred from Marketing to Development to proceed with the approved design prototyping.
Gate Review 2—When the pre-production product development is ready for manufacturing ramp.
Gate Review 3—When the completed product is about to be offered for general sale.
Gate Review 4—6–12 months after Gate 3 checkpoint, when product support and acceptance is to be assessed, transfer of design authority to manufacturing evaluated and ultimate product disposal plans reviewed.

THE NEW PRODUCT MODULE

The new product module (NPM) plays a fundamental role in the gate process. The NPM operates as "factory within a factory," controlling its own manufacture of new PCBs and interfacing with diverse groups, both within and outside the division. The scope of this operation requires a multidisciplinary approach that relies on effective teamwork.

The NPM consists of the following interactive groups:

- Manufacturing engineers and project managers.
- A dedicated materials and planning group.

- A prototype shop within the main plant.
- A process development group.

The role of the NPM includes:

- Designing manufacturability into new products.
- Interacting directly with new product designers.
- Assessing the impact of design features, changes, and alternatives on manufacturing.
- Providing manufacturing engineering support to the new product module prototype shop.

The group consists of the following:

Elements	Roles
Manufacturing engineers and managers	Designing manufacturability into products
A dedicated materials and planning group	Sourcing and supplying new components, scheduling prototype builds
A prototype shop	Providing support to prototype shop
A process development group	NPI continuous improvement

THE MANUFACTURABILITY ASSESSMENT

Northern Telecom's corporate structure includes both the functional and physical separation of design and manufacturing operations. Bell Northern Research is the company's R&D arm, largely centered in Ontario, Canada. Its priorities rest mainly with basic research and the design and development of new products, including technology development for connectors, custom silicon integrated circuits, and complete networks. Northern Telecom manufacturing facilities are located throughout the world, with a major operation called the Bramalea works located near Toronto, Ontario.

The manufacturability assessment tool addresses the issues of manufacturability at the different company production sites. The assessment of a new design is a progressive, interactive process that uses on-line electronic design files in concert with design guidelines built into a corporate CAD tool.

For printed circuit boards (PCBs), the manufacturability assessment (MA) is a spreadsheet-based checklist that has evolved in conjunction with corporate and divisional standards that establish the design rules for Northern Telecom switching products. Its inputs are developed by the new product modules at different Northern Telecom manufacturing sites.

This assessment represents Northern Telecom's latest requirements for new product manufacturability. It is used by the NPM engineers as a tool to determine the manufacturability level of a new product by assigning an indexed rating to items or design features that do not conform to the manufacturing requirements at the production sites. The MA is conducted jointly with the design engineers at the early stage of the design cycle, to provide a measure of attaining project manufacturability goals and to provide vital manufacturability feedback to the design engineers.

Following gate 1 in the development process, the NPM engineers will access electronic design data files, to examine the layout of the PCBs and the associated component list. The components will be checked against several requirements, including the layout requirements, the placement locations, the preference level (generated from historical data based on the specification of the components and their availability), and any particular assembly requirements.

At this early stage, the MA is an interim process to provide feedback, going directly on line to the design engineers before any further investment into the product, such as phototools and the scheduling of prototype builds. The NPM team will then ensure that the design is revised to reflect their manufacturability inputs, before prototype or preproduction builds are scheduled and materials are ordered.

The full MA is conducted with the first prototype builds. Problems with actual assembly, solderability, in-circuit test, and repairability are recorded and documented by the NPM engineers in the MA (see Figure 8-2). This feedback is a vital element of this assessment. In addition, the summation of all design feature index scores is determined, and, if it exceeds a predetermined maximum, the design must be recycled (redesigned) to obtain a better score.

After the MA, if all the requirements are met, the design is allowed to pass gate 2 and proceed toward the transfer into volume manufacturing.

Manufacturability Assessment Criteria

The MA consists of several evaluation criteria used in the evaluation of PCB designs for new products (see Figure 8-3). These are:

1. PCB design.
2. Components.
3. Manual operations.
4. Testability.
5. Repairability.
6. Documentation.

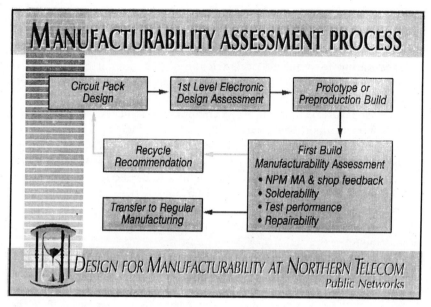

FIGURE 8-2. Manufacturability assessment process.

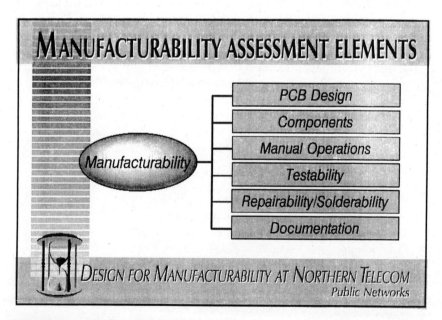

FIGURE 8-3. Manufacturability assessment elements.

PCB Design

Northern Telecom and Bell Northern Research have developed an interactive CAD system for PCB design and analysis. Many complex PCB features are created, and the assessment of the different parameters involved in the design, such as hole and land placement and dimensions, are related to subsequent fabrication and assembly process quality, such as the insertion yields of through-hole components and their solderability. Other factors, such as component body dimensions, lead diameter, component orientation, lead forming, PCB marking, and body clearances are evaluated during the MA. Special attention is given to track and via routing. Specifications are given for SMT and mixed technology components and their locations in the PCB layout (See Figures 8-4 and 8-5). The PCB design system output is also examined for clear presentation and markings, to reduce future assembly errors.

Components

For through-hole technology, components are selected for automatic insertion success. Component body diameter, lead diameter, and span must fall within the documented manufacturing process capability (see Figure 8-6). Labor-intensive configurations, such as vertical mounting, off-grid, and diagonal placements, are unacceptable and must be avoided. Failure to do so will result in a poor manufacturability index score, and may cause the PCB to be redesigned.

Manual Operations

The manufacturability assessment will encourage new product designers to minimize the proportion of components requiring special assembly operations. The evolution in packaging technology is aiding in the sharp reduction in manual assembly operations. Nonstandard components, such as those requiring heat sink assemblies or magnetics or those that are back loaded after soldering because they cannot survive the standard assembly process, are being eliminated or redesigned as much as possible.

Testability

Test engineers are part of the NPM team and generate testability guidelines and checklists. There are designed-in testability requirements such as the physical test fixture access to test points on the PCB and other features, such as tri-state logic, peripheral, and boundary scan testing. A separate testability assessment checklist is used to verify that these essential requirements are met.

144   Successful Implementation of Concurrent Engineering

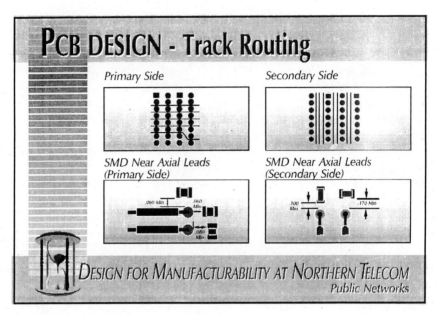

FIGURE 8-4.   PCB design-track routing.

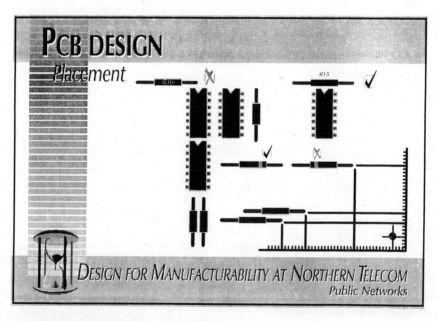

FIGURE 8-5.   PCB design.

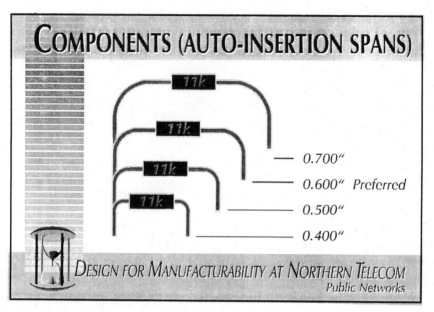

FIGURE 8-6. Components (auto-insertion spans).

Repairability
As product technology evolves, future improvements may require that PCBs be returned to the factory for repair or update. Changes in the component technology may make such repairs or updates at the highest quality difficult and costly. Repairability guidelines have to be applied during the manufacturability assessment, to evaluate the future complexity and cost of repairs. This is especially congruent with the growth of multilayer PCB and SMT technology, including the continued evolution of fine-pitch SMT devices.

Documentation
Design, manufacturing, field, and customer documentation must be in place to support manufacturing and volume product shipments. The development, specification, and availability of such documentation is an important element of the MA and is a prerequisite of the gating process.

THE EVOLUTION OF THE CORPORATE STANDARDS

The ongoing evolution of a set of corporate standards to guide all future product development is an important part of integrating and standardizing the design and manufacturing information throughout the company. In ad-

FIGURE 8-7. Corporate standards.

dition, these standards eliminate the duplication of engineering efforts, improve communication in the engineering and manufacturing communities, accommodate legal and regulatory requirements, and lead to enhanced quality and manufacturability. These regulatory requirements include ISO and UL/CSA standards (see Figure 8-7).

The Northern Telecom corporate standards are embodied in the software CAD design tools to verify and check all design data. The design system thus becomes the vehicle to apply the corporate standards and enforce the manufacturability rules.

The NPM engineers are instrumental in the development and evolution of corporate standards. They provide the link between the changing manufacturing environments at the different production locations and the corporate standards, to ensure that the design tools are as up to date as the products that they were designed to create.

RESULTS OF THE DESIGN FOR MANUFACTURABILITY EFFORT AT NORTHERN TELECOM

Northern Telecom has developed a successful strategy for new product development in order to react quickly to market demands for the highest

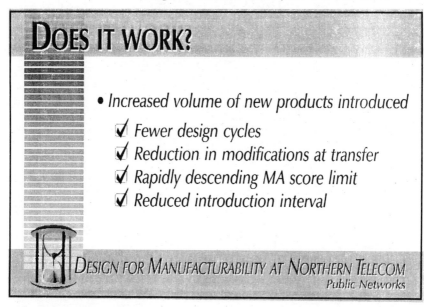

FIGURE 8-8. Does it work?

quality, a low cost, and time-to-market (see Figure 8-8). Dramatic results have been obtained from this process, including a substantial increase in new product introduction. In the period 1986 to 1991, the following have been realized (see Figure 8-9):

Average design cycles for PCBs	60% reduction
PCB introduction interval	75% shorter
Number of modifications at transfer	100% reduction
Average manufacturability assessment	88% lower

THE FUTURE CHALLENGE

It is interesting to note that the average manufacturability scores have been increasing since 1990. This is due to the change in design requirements, which leads to increased demands on the manufacturing process. Examples would be the application of more sophisticated components, such as VLSI and fine line SMT devices, causing increased PCB layout density, down to 25 mil lead pitch.

This trend reinforces the importance of continuous improvements and using the tools and elements of concurrent engineering, so that the optimum balance between the design and manufacturing requirements can be

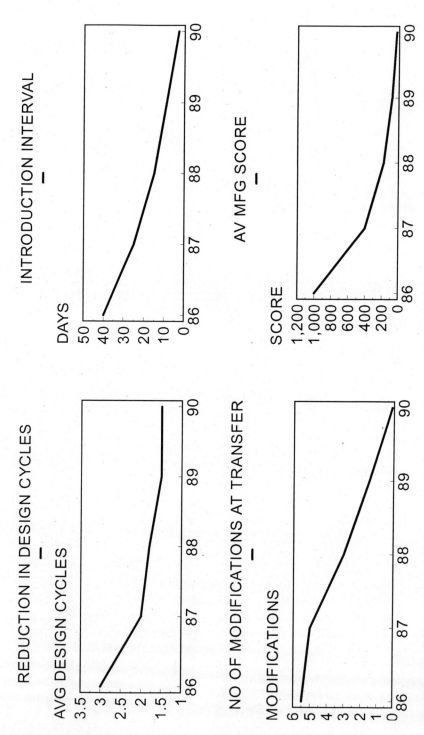

FIGURE 8-9. Northern Telecom Digital Switching Division new product development.

achieved in the shortest amount of research and effort. Northern Telecom's new product development process is answering this challenge through developments in assembly simulation, mechanical standardization, expert systems, and experimental design.

9
Concurrent Engineering of a Polaroid Camera and Flexible Automation

George Reimann
Polaroid Corporation

THE NEED FOR CONCURRENT ENGINEERING

Product life spans are becoming shorter than in previous periods. Rarely does a typical product have a 5–10 year life any longer, but more typically a 2–3 year life. This demands that companies like Polaroid must reduce the time to get new products to market. No longer can the approach of engineering the product prior to the engineering of the assembly equipment be tolerated. This approach puts both tasks in series. Feedback on product improvements takes too long, and the entire time to reach production automation goals takes too long. The present markets for products are no longer local only, but typically are global in nature. Thus, competition comes from all countries to be the first to market a new product.

Concurrent engineering is the process of developing the design of the product at the same time the automation equipment is designed and developed. Concurrent engineering makes use of all the latest tools available to speed up the process: computer-aided design (CAD), computer-aided engineering (CAE), design for manufacturability (DFM), process capability (Cp), computer simulations of automation, and, in Polaroid's case, flexible automation equipment.

Although it gets the product to market more timely, concurrent engineer-

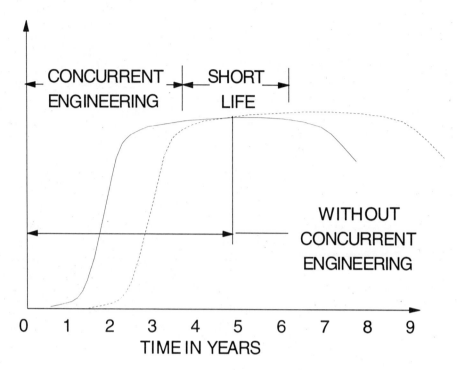

FIGURE 9-1. Relationship of concurrent engineering process vs. time.

ing carries some inherent cost penalties. As the product develops, original design tooling may have to be scrapped and new tooling designed.

THE APPROACH USING FLEXIBLE AUTOMATION

Polaroid has chosen to use the *flexible automation approach* for its subassemblies. This concept was developed in Japan by Sony Corporation and was first used in their Walkman production. It consists of *robotic assembly stations* (RAS) incorporated into a complete parts delivery system. The system concept is called a SMART™ system by Sony (Sony Multi Assembly Robot Technology). The parts delivery consists of two types of delivery systems. One, for bulk parts delivery (the preferred system, as costs are the lowest for parts handling), is called the *automated parts orientation system* (APOS). This system is the key to the success of small parts handling, as one APOS machine can deliver up to six different parts to the RAS for assembly. The other type of parts' handling is *automated tray changer* (ATC), which takes parts that are too large for the APOS concept and delivers trays loaded at the piece part vendors to the RAS for assembly. The connection of several RAS cells, an APOS, and ATCs make up the complete SMART™ system concept. The assembly of the product is done on a tooling plate called a platen, which is presented to each of the RAS cells in the assembly process. All are easily retooled for new product.

For a completed product a station in the SMART™ system is called an *automatic parts traying* (APT) unit. This unit takes the product from the assembly platen and puts it into a tester and then into trays that are taken to the final assembly of the product. Polaroid camera typical subassemblies are shown in Figure 9-2.

This system is considered by Sony and Polaroid to be among the most flexible available because the system is primarily software driven. To change a product, usually only software changes are required. The actual hardware of the system that may get changed is only the APOS pallets (typically, only four are required per piece part), the RAS cell robot end effectors, trays, and the assembly tooling, called platens. The method of connecting the RAS cells and components allows one to easily reconfigure

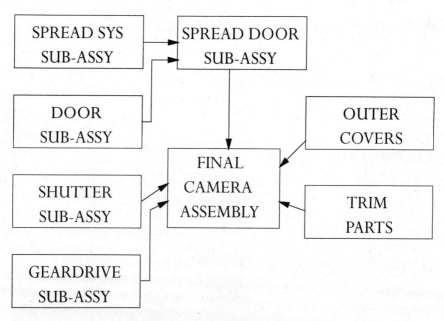

FIGURE 9-2. Polaroid cameras typical subassemblies.

Concurrent Engineering of a Polaroid Camera and Flexible Automation

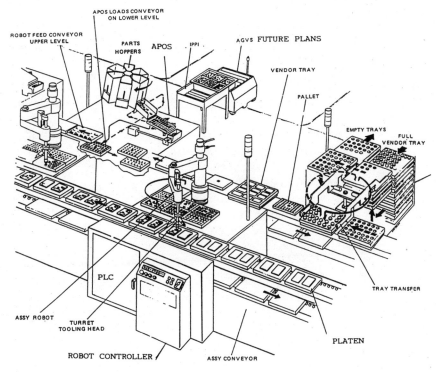

FIGURE 9-3. SMART™ flexible automation concept.

portions of the assembly line, should a product change take place that dictates it. Figure 9-3 shows the SMART™ concept of *flexible automation.*

PRELIMINARY ENGINEERING FOR FLEXIBLE AUTOMATION

In order to be able to get into a concurrent engineering mode for a new camera product, Polaroid entered into a learning phase that would allow its engineers to design the required tooling to assemble the product as the product was being designed itself.

All product designers were sent to CAD and DFM courses, along with Polaroid's own internal set of guidelines for DFM with flexible automation. Interactive teams were established between the product designers and machine designers on each subassembly. The entire product was designed on CAD, with full database access by the machine engineers and designers.

What Polaroid learned from the tooling design of the SMART™ system lines in production on the older products was categorized and put into library formats for all to review and reference. These areas of tooling are explained below:

1. Platen—The SMART™ line assembly fixtures that travel between RAS stations for assembly of the product. Typically, 30 units are needed for full production output.
2. Pallet—The acrylic piece part holders that are numerically controlled (N/C) machined, using the product data base and are loaded with parts by cycling through the APOS machines. A CAD pallet library was established for reference for new parts.
3. End effectors—The robot has a turret that can carry six end effectors (one for pallet and tray change and up to five for product parts or functions). These have now been standardized into eight categories of types for different product parts or functions.
4. Trays—These are used to bring parts directly from the supplier to the SMART™ line for assembly. They are used only if the part cannot be APOS loaded into a pallet.
5. Other tooling—The SMART™ system is also flexible in that other types of part supply tooling can be interspersed into the system. More traditional part feed systems are vibratory bowls, tube feeders, and magazines. It can also be married to *hard* automation machinery as a pre-assembly.
6. Computer simulation—A complete computer simulation was developed by Polaroid using SLAM and TESS. This simulation has been shown to be invaluable in the fine tuning of the production volumes. It shows all the interrelationships of RAS, APOS, ATC, stops, fill rates of APOS pallets, and the traffic control of the parts feeding conveyer.

SUBASSEMBLY AUTOMATION FOR THE NEW CAMERA LINE

Now that the automation of the present product lines had proven to be a viable concept, the plans were put into place to automate the new camera line. Its subassemblies are the same as all previous camera lines: a spreader assembly, a door assembly to house the spreader, a shutter assembly, and a geardrive assembly. A team was set up between the Product Design Group and the Computer Integrated Manufacturing (CIM) group, in order to assure that the product was designed for automation and that all assembly problems have a direct feedback to the Product Design Group. The same CAD data base used to design the product was used to design the lines and tooling for the flexible automation.

Figure 9-4 shows the interrelationships between functions.

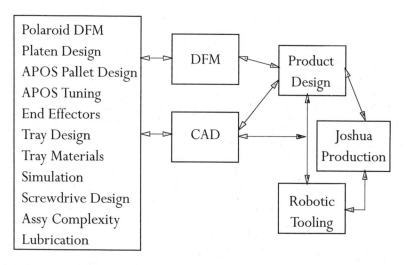

FIGURE 9-4. Tooling design and production relationships.

Spreader Subassembly

Because the heart of the Polaroid camera processing is the *spreader assembly*, it was determined that this would not be done on the SMART™ flexible system, but would be produced on a single hybrid hard automation/robotic machine. This line would produce 100% of the worldwide spreader systems for all models of the camera line. Therefore, its throughput rate would have to have an order of magnitude greater than any of the individual subassemblies, which would be produced on the SMART™ flexible automation system. The assumption—since the *spread system* is so critical to the picture quality—is that, once developed, it would not be subject to many design changes for different camera models. The *spread assembly machine* is a combination of seven robots for the material handling and assembly, walking beam transfers, vibratory bowl feeders, and trays. Concurrent engineering has improved not only the product's ease of assembly and quality, but also the assembly automation.

Figure 9-5 shows the spreader assembly machine.

Door Subassembly SMART™ Flexible Automation Line

The door is assembled utilizing the SMART™ flexible automation concept, utilizing six assembly robots and one unload robot. The parts are delivered

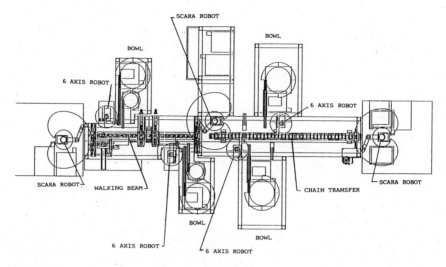

FIGURE 9-5. Camera spreader assembly machine.

to the robots using four types of pallets and 5 types of trays. Upon completion of assembly, the platen returns to the unload robot, where it is exercised, tested, printed with production data, and loaded into a tray by its robot, which then goes to final assembly.

See Figure 9-6.

Shutter Subassembly SMART™ Flexible Automation Line

The shutter is also assembled utilizing the SMART™ flexible automation concept. It is a longer, more complex line than the door line. It consists of sixteen assembly robots, two tester robots, and one unload robot. The parts are delivered in ten types of APOS pallets, six types of trays, two magazine feeders, two vibratory bowls, and two lens tube feeder units. The test loop for the shutter is very complex. The product is matrixed through two test stations that consist of three testers each. This is necessary, since the shutters are not only tested, but are also calibrated, taking more time than a single tester could handle for the throughput rate. Each shutter is tracked from the tester and identified, so that at the unload station a complete history is carried on it via a bar code. The unload station only unloads into trays, which then go on the final assembly of the camera. The shutter line is protected against static discharge due to the CMOS components. See Figure 9-7.

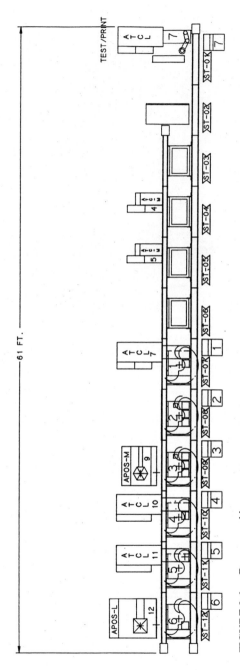

FIGURE 9-6. Door assembly process.

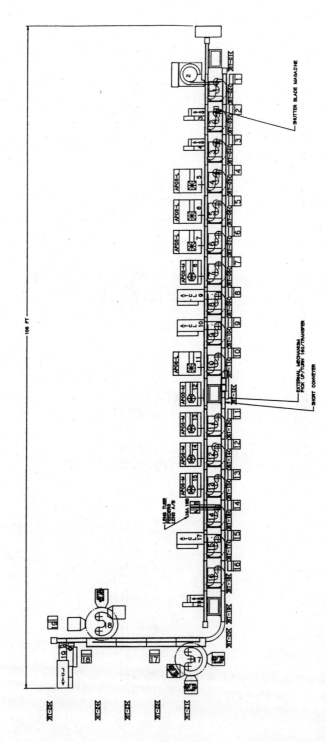

FIGURE 9-7. Shutter assembly process.

Geardrive Subassembly Flexible Automation

The geardrive SMART™ assembly line is several levels more complex than the shutter line. It consists of a *hard automation* section that robotically unloads the geardrive and an indexing turret that installs motors, screws, pins, and lubrication. Then the geardrive is transferred by a robot to the SMART™ assembly line for the balance of the assembly. The SMART™ assembly line is U-shaped due to its length. It consists of 22 assembly robots, 1 test, print, and unload robot, 20 types of parts delivered in APOS pallets, 7 trays, and 5 vibratory bowls. The geardrive line incorporates lubrication into the product in several areas, besides the geardrive itself. See Figure 9-8.

THE NEXT STEP—FINAL CAMERA AUTOMATION

At this time, the final assembly of cameras is by manual assembly, due to the large nature of the subassemblies and electrical interconnects that are performed at this time. Polaroid is developing concepts beyond the normal hard automation concepts that may allow final camera assembly to be automated in the future. This is presently a joint Polaroid and Sony Corporation of Tokyo, Japan, effort. See Figure 9-9.

SUMMARY

As a summary of the use of concurrent engineering for any product design, the advantages and disadvantages involved should be pointed out. As stated in the first part of this chapter, it is essential to get the product to market first. Concurrent engineering does assist this goal, but with a cost penalty, usually due to the effect of piece part design changes as the product design evolves. However, the upside of this cost penalty is that the product will be better engineered for function and ease of assembly. Once the product is in production, the production volume scale-up is an order of magnitude better than if concurrent engineering was not employed.

Table 9-1 lists the advantages and disadvantages of concurrent engineering, SMART™ flexible automation, and hard automation.

It becomes apparent, from Table 9-1, that computer-aided design, design for manufacturing, and design for automation are all viable tools of concurrent engineering, regardless of the manufacturing method planned. These

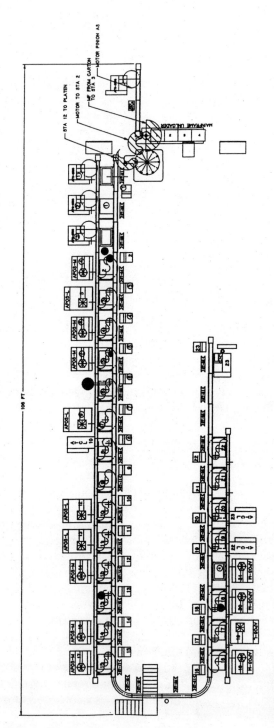

FIGURE 9-8. Gear drive assembly process.

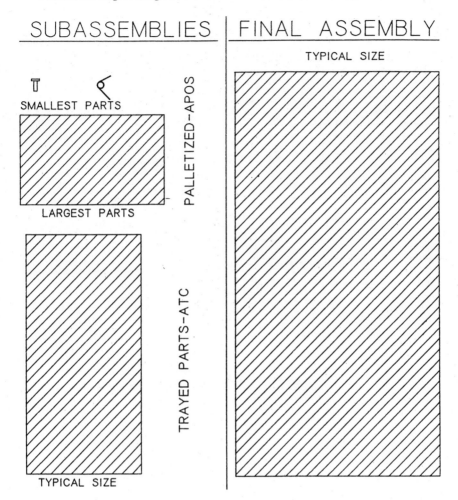

FIGURE 9-9. Relative size of subassemblies and final assembly.

tools will result in a more easily assembled unit, even if manual assembly of the unit is employed, and will attain production rates in the least time. In attaining any equipment ROI (capital justification), one must keep in mind the fact that manual assembly will be assisted by the concurrent engineering process.

However, the quality improvements will not be realized unless the repeatability of robotic assembly or hard automation is employed. Polaroid's experience has proven that the automation of a product versus manual will make a measurable improvement in the product quality.

TABLE 9-1. Concurrent Engineering

(Applicability to Manufacturing Methods)

CE tool or feature	SMART™ flexible automation system	Hard automation system	Manual assembly
Computer-aided design	Yes	Yes	Yes
Design for manufacture	Yes	Yes	Yes
Design for automation	Yes	Yes	No
Assy computer simulation	Yes	No	No
Software driven	Yes	Some	No
Cp	Yes	Yes	Yes
Engineering cost	Medium	High	Low
Hardware cost	Medium	High	Low
Tooling cost	Medium	High	Low
Sensitivity to part change	Low	High	Low
Sensitivity to assy order	Low	High	Low
Time to get to production	Short	Medium	Short
Cost for retooling	Low	High	Medium
System throughput	Medium	High	Low

Polaroid's experience has typically shown a cost penalty risk of less than 5% with concurrent engineering and SMART™ flexible automation, for first time projects. As concurrent engineering tools and processes mature, Polaroid believes that the penalty risk can be cut to less than 2.5%.

10

JITQC: An Employee-Centered Methodology Designed to Impact Product Quality in an SMT Manufacturing Line

Jim J. Zebrowski

Digital Equipment Corporation

INTRODUCTION

In today's volatile marketplace, corporations are struggling to find proven recipes to remain competitive for the long haul. The stimulus for this can be found in the realignment of economies in light of the slowdown of consumer spending and the overall adjustment required for business to present its image to a more global market. As a result, managers are scrambling to streamline and realign their manufacturing lines, to make them more flexible and cost efficient.

This chapter illustrates how one small electronics manufacturing line at Digital Equipment Corporation started the process to make practical adjustments toward meeting the performance goals specified by corporate management. Every major corporate goal and objective had to be translated into practical evolutionary common sense by individual managers, supervisors, engineers, technicians, and manufacturing operators throughout the company. Above all, success can be guaranteed only by the involvement of all discipline levels that can make it happen for the manufacturing line.

A commitment to meeting corporate goals today is a commitment to meeting change head on. The strategy that this electronics manufacturing line used to meet corporate goals was to align itself with a policy that promotes a program of continuous improvement. Areas that were considered an essential ingredient in defining improved product quality were highlighted, and programs were introduced to track metrics that reported on performance issues in these areas.

The most practical way for manufacturing to meet performance goals was to implement a series of incremental process improvements that affected the product quality in this electronic production line. The issue of becoming proactive in defining what your customers' needs are now and in the future played a critical role in defining this manufacturing line's new approach to business. Management's role was also critical in emphasizing the importance of employee motivation in bringing about this change.

It is important for companies embarking on new product quality improvement programs to encourage success by carefully orchestrating their introduction. Widespread acceptance and confidence can be nurtured by targeting specific product lines as part of a gradual phase-in process for any change in manufacturing methodology. Management can also assist by adjusting their commit schedules to coincide with the new way the manufacturing line actually works to meet corporate objectives.

These goals can be addressed through long-term planning that values the pursuit of excellence through continuous incremental improvements in each assembly process. This idea allowed the manufacturing line to aggressively attack issues that most impacted product quality by focusing their strategy in three areas. The areas included new tooling geared to improve the process capability for electronic packages, the introduction of a *just in time with total quality control* (JITQC) through employee involvement, and the solicitation of customer feedback on how they perceive the process.

The role of communication was essential to the success of this process because it was used to elicit information and to inform the corporation and customer of progress. This progress was seen as beneficial to both and as a stimulus for broadening the technical capabilities of the manufacturing group. Customer feedback also allowed the manufacturing line to focus on the benefits most valued by the customer.

The concept that drove the introduction of JITQC through employee involvement was the need to reduce *work in process* (WIP), improve yields (reduce PPM defect levels), and lower *cycle time*. The improvements in these areas highlighted issues that normally prove difficult to resolve because poor management of these factors can mask major process problems, as manufacturing focuses on just meeting the numbers. The implementation of a JITQC process would allow engineering to concentrate on correcting tooling deficiencies by collecting weekly defect data. Once the defect code data was quantified into categories, engineering could assess exactly how well the manufacturing process was running.

This information could now be correlated with customer supplied feedback, so engineering was able to decide where to maximize its effort for the best value-added benefit. At this point, it became critical for process engineering to work closely with manufacturing to decipher the root cause for each of the largest contributors to product defects. This is where a multidisciplined approach to problem solving paid big dividends. Often, the manufacturing engineer only sees a portion of any problem, whereas the operators and equipment technicians provide valuable insight due to their longer daily manufacturing exposure.

A final critical area was to focus the entire manufacturing line on problem solving geared to defect avoidance. Only a real time report–generating tool managed by all involved parties provided the immediate reinforcement needed to address issues as they arose rather than weeks later. Management backing was instrumental in providing the resources necessary to develop and support the installation of this on-line data collection system, geared to statistical process control (SPC) charting.

METHODOLOGIES

The manufacturing line involved in this study provided final testing and assembly processing to gull-winged electronic surface mount (SMT) packages that were mounted on printed wire boards inside a large computer. These packages housed the silicon semiconductor chips that communicated their functions through the external leads surrounding the outside of each package. Manufacturing operations, such as test, heat sink attach, package lead forming, lead tinning, and final dispatch in customized shipping media, were performed by this group.

There were more than a dozen different types of packages that were processed through this line by documented procedures that held each to a customer's specified tolerances. All assembly steps in this line were under SPC, and technicians responded immediately to computer-generated control charts if any process drifted out of control.

The tolerances mentioned were incorporated into manufacturing floor documentation that allowed technicians to produce electronic packages that aligned precisely on customer-printed wire boards. These values were specified around a nominal number, and all equipment was set up to produce product within a plus or minus value around this nominal number. These values were reflections of the customer's own manufacturing process and allowed the product to be easily handled by their automated vision placement equipment.

The actual measurement data reviewed in this study was captured from an SPC and a corporate work flow data base used by the operators in this manufacturing line. This routine data gathered by manufacturing technicians using a programmable robotic vision machine captured lead dimensional tolerances, such as coplanarity mean (vertical displacement of the package leads relative to the seating plane of the package), skew mean (side-to-side displacement of the leads), and lead tip average one (actual distance measured from the lead on one side of the package to the lead on the other side). In each case, the mean recorded the value of the worst case high values for the two sample packages evaluated for SPC data.

Survey data was also recorded to evaluate the perceptions of manufacturing operators before and after the introduction of JITQC. This provided a method of quantifying the impact of employee involvement on the implementation of JITQC. Another set of data was provided by a corporate engineering task team that visited customer sites and analyzed their sensitivity to all the process improvements made by the manufacturing line. The study showed a steady rise in product quality, as defined by the three measurement categories.

Since the line was under SPC, the data entry process was determined by a specification that was regulated by the document control department. Data input was made through a computer into a data base that provided a real-time view of the SPC charts. This allowed each manufacturing technician to make any process adjustments needed if any SPC rules were violated.

The three main areas and the methodologies used to determine their impact on customer product quality were:

1. New tooling introduction—New tooling in the form of die sets was introduced to gain control over customer-specified tolerances. This tooling ran on hydraulic presses that used different die sets to form (shape) the package leads for placement on printed wire boards (PWBs), according to package type and the customer specifications. The actual tooling studied was the 172 package die set, but earlier experiments with other package styles helped focus the development process on this package.
2. JITQC through employee involvement—This concept minimized work in process (WIP), helped define efforts at defect reduction, improved cycle time, and allowed quicker response to any deviations in SPC. The process of implementing JITQC was accomplished through the systematic efforts and cooperation of all employees, including management. Employee involvement was an essential ingredient integrated into the way JITQC was introduced to the people in the manufacturing line.
3. Customer feedback—Data provided by the customer was reviewed to determine if the incremental improvements made by the manufacturing line had the intended impact on product quality at the customer's site. There was also some descriptive information that indicated if the customers perceived that progress was made in addressing their needs.

Populations

The population for the statistics gathered in the study of new tooling involved sampling all the packages processed through manufacturing for the entire life of the 172 LDCC device. Comparisons were made of the data gathered at the same location by the various iterations of die sets, to see if any improvement was noted. The study used the SPC sampling unit of 2 out of every 20 ($2/20$) packages processed through the area reviewed for data analysis.

This number was established by first running a process capability study with fixed parameters for a period of 2 weeks at a sampling rate of $2/5$ packages. (A process capability is done to determine the upper and lower SPC control limits, based on a fixed optimized set of parameters.) The SPC data was reviewed, and then every other value was removed to see if there was any significant loss of important data (mean, standard deviation, etc.). When no loss was seen, the capability study was run again and reviewed in the same manner until the sampling rate of $2/20$ packages was determined to be the optimized SPC sampling rate needed to capture essential data, with no significant loss of information.

Research Design and Procedure

The data gathered during the installation of the new tooling die set was input into a computer table by line technicians, using the manufacturing sampling number of $2/20$ packages that were run. A comparison between the old and new die sets was displayed in a bar chart form to illustrate the changes that different die sets had on the process.

Information from historical data on the corporate data base and employee survey input was used to show how the new JITQC principles impacted product quality. The analytical information from the customer site represented data gathered by a corporate engineering task team interested in developing a standardized customer metrology base for measuring product quality. This data was displayed to show the impact of the incremental changes made by the manufacturing line with the introduction of new tooling and JITQC through employee involvement.

The data from the customer site provided skew improvement values that directly addressed how successfully packages can be placed on their printed wire boards. They also provided bar charts, showing the package data before and after a period of time for different types of packages shipped from this electronics manufacturing line. This was also viewed as an ideal opportunity to generate feedback from a customer that could optimize manufacturing process improvements for benefits most valued by the customer.

Statistics

The statistical analysis of the data was handled by using software called RS/1 Graph and Statistics Program from BBN Software Product Corporation, Cambridge, Massachusetts, to calculate and interpret all data collected and displayed.

Tooling

Customer specifications for coplanarity mean, skew mean, and average lead tip 1 were compared before and after each new die set went on line. This data was subjected to a variety of different tests, including a group analysis of variance (ANOVA) and multiple comparison tests, to determine if there was any statistically significant differences at the 5% probability level. Statistical information, such as the mean, standard deviation, range, and upper and lower control limits, were displayed.

Customers required that leads be formed within a narrow set of tolerances measured in thousandths of an inch (.001 inch), so their assembly process robot could mount the electronic packages on printed wire boards. A difference of a few thousandths could easily cause major placement difficulties on boards that had a placement target area of only .016 inches.

Data Collection Process

A computer program displayed a set of SPC charts that allowed the manufacturing technicians to proceed with their work or shut it down if the program determined that any data entry point was out of SPC. The values that were selected for this study were the coplanarity mean, skew mean, and the lead tip average on one package, as measured by the manufacturing technicians and entered into this data base. The time period for this study started in February 1991 and continued until November 1991.

JITQC and Employee Involvement

Measurement Data: Data Collection Process

The second category investigated was the impact of JITQC through employee involvement on the manufacturing line. Here, two distinct sets of data were collected. The first set involved measurement data that showed trends before and after the introduction of JITQC to the line. This measurement data covered the areas of WIP, yield, and cycle time and represented data that was routinely collected and evaluated on a weekly basis. The improvements in this part of the study were demonstrated statistically by subjecting the numerical measurement data to the Mann-Whitney test for differences in the medians.

JITQC Survey Data: Data Collection Process

The second set of data was collected through a survey that tried to measure the level of employee involvement in manufacturing activities. JITQC by itself could not be successfully introduced without the close cooperation of all the manufacturing technical people. The survey tried to evaluate the perceived changes that employees felt occurred before and after the introduction of JITQC to the manufacturing line.

The survey was composed of two parts, with a set of nine statements that were repeated exactly in each part. The survey takers were asked to respond to a series of statements and to record their feelings to each statement by checking an evaluation category. There were five evaluation categories, ranging from strongly agree on one end to strongly disagree on the other end. There were also two brief demographic questions in the beginning and a final question in the second part that allowed the individual to enter a response in his or her own words.

The survey results were organized by assigning each response with a number and then placing it in a data base for evaluation. The data was then displayed in bar chart form, to see the changes before and after the start of JITQC. Descriptive statistics were displayed in table form, and an assessment of the data was given.

Customer Feedback: Data Collection Process

The final category involved information gathered at customer sites that revealed their perceptions of any improvement in product quality. The intention was to relate any incremental improvements made by the manufacturing line with new tooling and JITQC to an actual rise in product quality noted at the various customer sites. This took the form of data collected at the customer site in a joint study sponsored by a corporate engineering task force with the full cooperation of the various customer sites.

The charter of this engineering task force aimed at establishing a universal metrology capability at each customer site so that they could react to and fully understand any shifts in product quality levels. This task force's initial role came in response to customer complaints about placement errors with a new electronic package.

This role expanded when the engineering group realized that most customer sites had no dedicated equipment devoted to quantifying incoming product quality. This made it difficult to differentiate incoming quality concerns from on-site process and equipment problems.

The engineering group devoted a major portion of its effort to developing a portable measurement system that could be calibrated against the automatic measurement system used by the electronic manufacturing line. They used this system to make accurate measurements at the customer sites, and then returned to analyze the data with a process specially developed to handle this information.

The data and analysis was published in a series of reports that allowed the electronics line and their customers to focus efforts at improvements in areas targeted for the maximum impact on product quality. This information also revealed that the manufacturing line had improved product quality over the period of time in question.

The customer site study was used to create some bar charts that displayed improvements witnessed at their assembly lines. This data displayed improved *worst case lead skew* measured in the course of the study from February 1991 to November 1991, plus it highlighted areas that required additional work. Efforts are still under way to improve data collection and define a metrology system that could be used by all customer sites. This will allow for more uniform monitoring of improvements, and it will help create a shared data base that can be accessed by all customer sites on a corporate-wide basis.

DATA ANALYSIS

New Tooling

The data base account was reviewed for information on the number of different die sets that were used over the life of the 172 LDCC package type. It was determined that useful data for this study could be gathered from February to November 1991 for the following responses: coplanarity mean, skew mean, and lead tip average for one package. During this time period, a total of four different die sets with two different subcategories were used in producing the 172 LDCC package. (The subcategory was an attempt to see if there were any equipment-related variations in the collected data. None were found. See Table 10-1 for a complete list of die sets.)

These die sets allowed the manufacturing line to handle variations in package styling and to try improvements to understand what design features had the greatest impact on product quality.

The four different die sets and the two subcategories to be studied are referred to by labels entered in the data base. They are as follows: 172F1_C2, 172F2_C2, 172FCD, 172FCD_C1, 172FCD_C2, and 172FLS. The 172 refers to the 172 LDCC ceramic package (a top-brazed, gold-leaded quad flat pack), and the other numbers reference the manufacturer of the die set and how it was used in line. (The FCD die set is the one with two sub categories—C1 and C2.) Table 10-1 represents a complete breakdown of the die sets, and Figure 10-1 and 10-2 show the initial handling of the data by comparing coplanarity mean (COPMN), skew mean (SKMN), and lead tip average one (LTAV1) data for the six selected die sets.

TABLE 10-1. Die Sets Reviewed in this Study and the Sample Package Populations

Die sets	Sample packages 2/20
172F1_C2	162
172F2_C2	106
172FCD	138
172FCD_C1	54
172FCD_C2	70
172FLS	160

A complete breakdown of the die sets in this study, with the 2/20 sample population used for each.

The four categories and two subcategories that were selected had population samples that ranged from a low of 54 entries to a high of 162. Each of these entries was made according to a 2/20 package sampling rate and was recorded on a daily basis by manufacturing line operators, who entered the data as required by procedures written and filed at document control. Based on the 2/20 sampling data, the total population represented by this study ranged from 1,080 packages to 3,240 packages.

The boxplots in Figures 10-1 and 10-2 allowed the data to be presented in a way that could easily determined differences between die sets and indicate the best way to proceed with the statistical portion of the analysis. The boxplots suggested major differences between the 172F1_C2 and 172F2_C2 die sets and the rest of the entries. From this initial look at the data, it was decided to use the Multiple Comparison test feature of the BBN software package RS/1.

The one-way analysis of variance (ANOVA), performed as part of the Multiple Comparison test, indicated that die set does have a statistically significant impact on each of the defined response categories of COPMN, SKMN, and LTAV1. The comparison showed that the means for the FCD, the two FCD subcategories, and the FLS die sets were not very different at the 95% confidence level. It did detect significant differences ($\alpha = .05$) between the 172F1_C2 and 172F2_C2 die sets and the rest of the die sets. (The die sets are the most recently introduced ones).

By using Dunnett's Multiple Comparison test based on the data discussed above, the information was further analyzed in order to define a quantifiable difference for each category. This test allowed comparisons to be made with a pre-selected category (172F1_C2 was selected as the control category referenced because it was the most recently improved die set) and to use this to show the statistical differences between it and all other categories. In every case, the means were shown to be different, the statistical value was recorded, and the degree of difference was highlighted as a number from 2 to 6, with 1 representing the control die set (the best value).

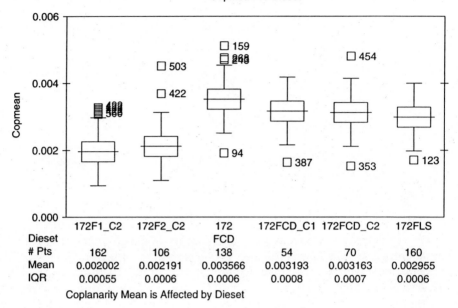

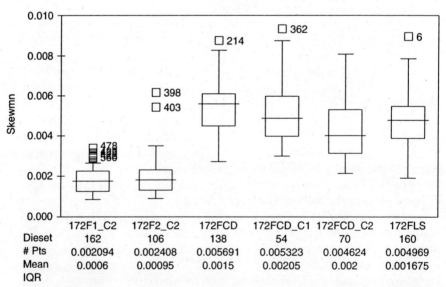

FIGURE 10-1. These bar charts show that the latest die sets offered significant improvements over earlier versions. (172F1_C2 and 172F2_C2 represent the latest versions.) Coplanarity mean and skew mean represent maximum numbers. The upper spec limit for coplanarity is .004 inches and .0037 inches for skew.)

JITQC: An Employee-Centered Methodology

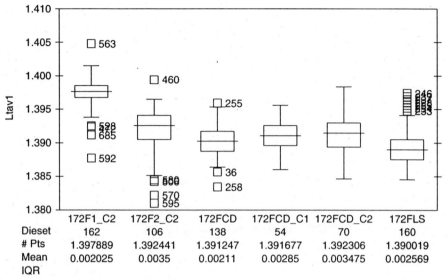

FIGURE 10-2. This bar chart shows that the latest die sets offered higher lead tip values over earlier versions. (172F1_C2 and 172F2_C2 represent the latest versions.) The customer reported improved coplanarity and skew control resulted in reducing the critical need for lower lead tip. Product end-of-life also factored in a decision not to bring this value under control. (The upper spec limit for LTAV1 was 1.397 inches.)

This data suggested that the latest iteration of the 172 die set family did offer significant improvements over earlier tooling versions, except for the response of LTAV1, which was waivered by the customer. When the latest 172 die set (172F1_C2) was released, the two responses of COPMN and SKMN were optimized and best approximation was used to improve LTAV1. The initial two responses improved significantly as expected, but the LTAV1 increased over earlier versions.

A better understanding of what controlled this response was determined, and moves were initiated to implement the required design changes. However, the customer site where this product was shipped intervened when it found that it could live with a higher level of LTAV1, based on the significant improvements made to COPMN and SKMN. They also did not want to fund this incremental improvement project when the product was facing an impending end-of-life cycle.

This end-of-life cycle was estimated to occur within 6 weeks of when the new die set could first be released to production. As a result, a permanent waiver was issued to allow LTAV1 to remain high for the remainder of this product's production run.

JITQC Through Employee Involvement

The data compiled in this category was comprised of a combination of hard measurement data and survey data used to map employee perceptions of involvement before and after the implementation of JITQC.

Measurement Data

The measurement data was taken from a materials tracking data base and was reviewed from July 1990 to December 1991, to assess the trends before and after the introduction of JITQC. JITQC was officially started in the manufacturing line during the month of April 1991 at the form/solder area that started the first JITQC kanban. The data comparisons used 38 weeks of entries prior to JITQC and 40 weeks of entries after JITQC for statistical analysis testing.

Improvements made by JITQC were highlighted by the dramatic drop in WIP and the improvements in yield through the reduction of defects. This was caused by setting up the JITQC kanban system that managed the flow of inventory on the production floor. When inventories decreased, there was a shorter response time to out-of-control product that reduced the parts per million (PPM) defect level and resulted in improved yields. Finally, cycle time decreased as yields improved, and the manufacturing line stopped going down for process-related issues.

The trends were displayed in a series of trend charts (see Figure 10-3), with the two sample populations compared before and after JITQC. Differences were highlighted, and a Mann-Whitney statistic was displayed for the categories of WIP, yield, and cycle time (Table 10-5). All three categories showed significant differences, and these differences revealed that WIP and cycle time decreased while yields improved as the ppm defect levels went down.

JITQC Survey

The second part of the JITQC category utilized a survey that tried to capture the level of involvement in improving manufacturing that employees had before and after the introduction of JITQC. The population for this study comprised all the manufacturing technicians, equipment technology technicians (ETG), process engineers, process technicians, supervisors, and trainers involved in the form/solder kanban area where JITQC was first introduced. This represented a total population of 24 people.

The response rate was 100% of the individuals attending the weekly meetings where the survey was administered. It was administered from

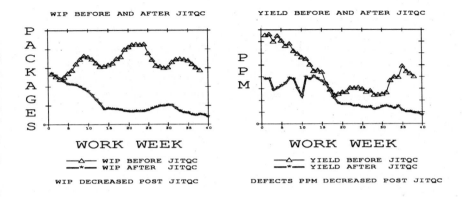
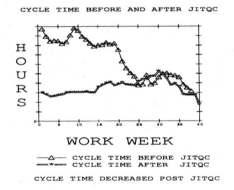

FIGURE 10-3. Work in process (WIP), yield, and cycle time improved after the implementation of JITQC.

December 18 to December 22, 1991, to accommodate all three shifts that worked in the manufacturing line. People were informed of the intent of the survey and were given a written as well as a verbal explanation of how to properly fill out the survey. (A sample copy of the survey is found in the back appendix attached to this study.) Additional explanations were given to individuals who use English as a second language in order to be sure they understood all the requirements.

The survey totals were created by assigning a value of one to each response category, from strongly agree to strongly disagree. Each question was evaluated for the total number of persons who answered in each category, and then the totals were added for all nine questions. This allowed a comparison of all response categories before and after the implementation of JITQC.

Survey results were displayed in a bar graph chart that used descriptive statistics to illustrate the level of responses in each category, from strongly agree to strongly disagree. The chart in Figure 10-4 shows the total level of change before and after the implementation of JITQC. The results show that the people responding in the Strongly Agree category almost doubled after JITQC (from 33 to 64). The totals in the Agree column also increased, from 82 to 110, after JITQC, while all other categories recorded a decrease. This showed an increased level of satisfaction with employee involvement in the manufacturing line after JITQC started.

The results for each of the individual statements are illustrated in Table 10-2, with all relevant statistics. The results from all questions showed an overall increase in satisfaction that there was a greater degree of employee involvement after JITQC. The individual response question at the end of the survey showed that, although things had improved, additional work needed to be done. These responses indicated that people wanted more participation in troubleshooting problems with engineering.

Customer Feedback

Customer site information showed an improvement over the course of the engineering task group study. Comparisons from March 1991 to September

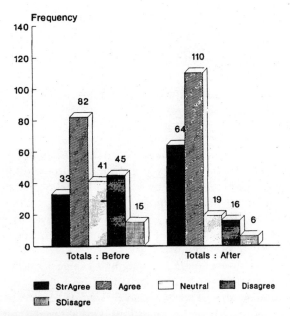

FIGURE 10-4. Survey results indicate a higher level of involvement after JITQC.

TABLE 10-2. The Survey Statistics Show Higher Employee Involvement After JITQC

Statistics	Before JITQC statements 1 to 9	After JITQC statements 1 to 9
Mean of all statements	3.386	3.979
Sum of all means	30.476	35.810
Median of all statements	3	4
Range of all statements	1.726	1.25

Survey Statistics
Scoring: 5 = Strongly Agree to 1 = Strongly Disagree

The survey statistics for after JITQC statements indicate a higher level of employee involvement in the manufacturing line, which supports the study's hypothesis.

1991 at customer site 1 showed skew defects of the 172 LDCC product reduced from 52% to 8%. (Only one customer site data was reviewed due to time constraints involved in this study.) Another product line called the LDCC 224 also showed a reduction in rejects due to skew, from 11% to 1.5%. (See Table 10-3 below.)

Another sampling of information from customer site 1 showed that the worst case lead skew (WCL skew) for the 172 LDCC package was well within the tolerances specified with only two outliers (see Figure 10-5). The population studied represented three different package lots, with the population distribution composed of packages with different foot lengths and housed in different shipping media as part of a continuous improvement experiment.

Only the outer three corner leads of each package were measured in an effort to speed results, but this was deemed appropriate, since these leads were the ones most susceptible to being out of tolerance. No significant differences were found in the three populations, which meant that the parts were received according to the requested customer tolerances.

Another major cooperative experiment run with 224 LDCC packages bound for customer site 1 showed that the lead skew changed during the shipping process (see Figure 10-6). This evaluation was performed on a

TABLE 10-3.   Customer Site 1 Feedback

March 1991	September 1991
Skew 52%	Skew 8%
172 LDCCs	172 LDCCs
Skew 11%	Skew 1.5%
224 LDCCs	224 LDCCs

This comparison study of incoming quality shows lower skew rejects after JITQC.

178   Successful Implementation of Concurrent Engineering

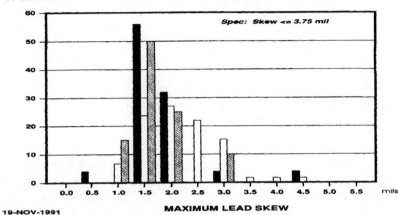

FIGURE 10-5.   Customer feedback on worst case skew distribution.

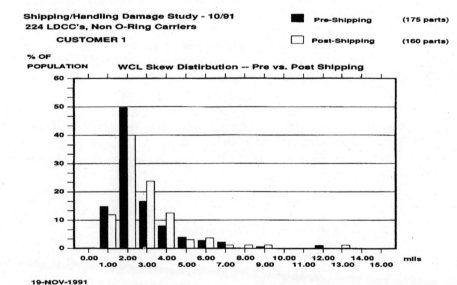

FIGURE 10-6.   Customer feedback on worst case skew for 224 LDCCs pre- vs. post-shipping.

population of 175 pre-shipping packages and 160 post-shipping packages. The findings illustrated in Figures 10-5 and 10-6 suggested that there was some shipping damage that caused the shift, so efforts were made to improve the standard shipping tray to one with O-rings that held the packages more securely.

ANALYSIS

New Tooling

The measurement responses used to analyze differences for new tooling were extracted from a large data base created to continuously monitor manufacturing process compatibility. This selected data was transferred to a smaller table for easier data manipulation for statistical analysis. The data extracted included the date of entry, the die set type, the coplanarity mean (COPMN), the skew mean (SKMN), and the lead tip average for one package (LTAV1).

This table contained 690 entries, entered from February 1991 to November 1991 by manufacturing technicians as a routine data collection procedure used on the manufacturing floor.

The initial analysis used a one-way analysis of variance because this allowed a comparison between all the different categorical factors (individual die sets) and each of the selected measurement responses. This gave an F statistic, which indicated that there was a statistically significant relationship on a global basis between the category and the selected responses of COPMN, SKMN, or LTAV1, with an $a = .05$. All measurement units for the responses COPMN and SKMN were in thousandths of an inch, while LTAV1 was computed in inches. (See Table 10-4 for the exact measurement ranges for each response.)

All measurements were made on calibrated automatic vision system, with a $2/20$ package sampling method for all packages running through the line. The population sampled was the total amount of 172 LDCC packages that were run in the manufacturing line between February 1991 and December 1991, which represented a total population of 13,800 packages.

Each one-way ANOVA showed that the category die set did influence each of the measurement responses by producing a very high F statistic value

TABLE 10-4. Range Limits for Each Measurement Response

Measurement response	Range limits (inches)
Coplanarity mean (COPMN)	+/−.0040
Skew mean (SKMN)	+/−.0375
Lead tip average 1 (LTAV1)	1.380–1.400

in every case with $\alpha = .05$. At this point, a multiple comparison test was selected to compare individual die sets against each other. In order to be as accurate as possible, each of the distributions was subjected to a Pearson's Chi-Squared Test for normality.

Since there were six die set categories and three responses, there were a total of eighteen distributions tested for normality. Pearson's test indicated that seven out of eighteen distributions were not normal, so this suggested the use of a transformation process to normalize the data. Data must be normalized to improve the accuracy of statistical testing by converting measurement values to the square root or log of the number.

Even when this transformation process was performed by changing all values to log to the base 10, four of the eighteen distributions still remained non-normal according to Pearson's Chi-Squared Test. These measurements did show an improvement in two of the distributions, while there was little change in the remaining two.

At this point, a decision was made to look at each non-normalized distribution's skewness and kurtosis measurement, to see how close each curve might be to a normal distribution. The skewness, which measures the degree of asymmetry of data around a sample mean, would show a value of 0 for a normalized plot. Kurtosis, which represents the degree of peakedness of a distribution and its proneness to outliers, would value normalized distributions at 0. Knowing this, a comparison found that skew and kurtosis in two distributions showed 66% improvement, while the remaining two categories showed very little change after the transformation.

The two distributions that did not change showed values less than one, so it was felt that it was still useful to proceed with a multiple comparisons test. It must also be pointed out that the whole purpose behind these tests was to demonstrate a statistically significant difference for changes introduced between recent die sets and the older versions, which showed that improvement was made. Improvement is defined as decreasing values of COPMN, SKMN, or LTAV1, which indicates that the process capability is becoming more robust in meeting product quality standards. This study demonstrated that the new die sets did produce results that reduced the means associated with COPMN and SKMN, which improved product quality and customer satisfaction.

Dunnett's multiple comparison test allowed the derivation of a statistic for each die set by referencing it off the die set that produced the best product quality. The die set selected as the reference or control was the 172F1_C2 die set because earlier testing showed it to have the lowest coplanarity mean and skew mean values. The software ran this test using the group summary table created earlier and produced a table with statistics (see Table 10-5).

TABLE 10-5. Dunnett's Multiple Comparison Test Results. This showed that the two most recently improved die sets (172F1_C2 & 172F2_C2) offered statistically significant improvements over the other die sets.

172F1_C2 vs other die sets alternate hypothesis	COPMN observed statistic	SKMN observed statistic	Range
172F1_C2<>172FCD	26.889	35.392	6
172F1_C2<>172FCD_C1	15.985	24.038	5
172F1_C2<>172FCD_C2	17.148	31.314	4
172F1_C2<>172FLS	19.048	22.092	3
172F1_C2<>172F2_C2	3.750	4.003	2

JITQC Though Employee Involvement

Measurement Data

Measurement data comprised of WIP, yield (PPM level), and cycle time records were subjected to a Mann-Whitney Test for unpaired samples. This test was performed to give a statistic that was independent of whether the distribution was normal or not. (See Table 10-6 for details.)

The test null hypothesis for each category suggested that the median of the post-JITQC population (sample 1 in Table 10-6) was greater than or equal to the before JITQC population (sample 2 in Table 10-5). This was statistically rejected at a significance level of 5% ($a = .05$).

Supporting the alternate hypothesis indicates that the medians of all three categories after the implementation of JITQC were less, which is exactly as expected if JITQC was working. Successful implementation of JITQC was supposed to drive all three of the indicators lower. The reduction of WIP allowed critical engineering efforts to remain focused on solving problems that impacted product quality.

TABLE 10-6. Mann-Whitney Test

Statistics	WIP (# of packages)	Yield (PPM)	Cycle Time (hours)
Post-JITQC: Sample 1			
Median	3396	8238.5	72
Number	40	40	40
Pre-JITQC: Sample 2			
Median	13105.5	20009.5	153.5
Number	38 $a = .05$	38	38
Mann-Whitney Statistic	1512	1273.5	1388.5

At a significance level of 5%, the null hypothesis that the median of Sample 1 is greater than or equal to that of Sample 2 can be rejected ($a = .05$).

Improvements in yield and a shorter cycle time resulted from the increased effort by engineering and management to resolve manufacturing problems formerly masked by a large inventory (WIP). Problems related to an out of control process were flagged quickly within the JITQC kanban, and the entire manufacturing support group responded to solve these problems. Time and persistence were the key elements that resulted in improved product quality. JITQC provided a window of opportunity for all these improvements to happen.

Survey Data
The survey data was displayed in the form of a bar chart, to illustrate differences before and after JITQC (Figure 10-4). A separate table totaling all the response categories used before and after the implementation of JITQC showed a shift in employees' perception of involvement after JITQC (see Table 10-2).

In order to make JITQC work, it was necessary to improve employee participation. The series of nine survey statements were structured to determine if the employees in the manufacturing line perceived an improved level of participation and communication after JITQC was implemented. Using descriptive statistics, it can be noted that the Strongly Agree response increased by 94% after JITQC, while the Agree response also noted an improvement of 34.5%.

Customer Feedback
No statistical testing was performed on the customer site information because of the lack of numerical measurement data. Descriptive results based on data accumulated by the engineering task force that developed a standardized measurement metrology system for all customer sites showed that improvement was noted by customer site 1.

This improvement was displayed in Table 10-3 in the form of reduced rejection of packages for out-of-tolerance measurements for skew. Figures 10-5 and 10-6 displayed information that allowed the manufacturing line to concentrate on improvements that benefited the customer.

Conclusion

This study postulated that product quality would be impacted through the combined efforts to improve die sets, to implement JITQC through employee involvement, and to receive customer feedback on the benefits of the changes made. The determinations indicate that the latest die set improvement (172F1_C2) significantly improves process capability over both the 172F2_C2 and the much earlier versions preceding it. Both the analysis of measurement data and survey returns suggest that JITQC had a significant impact on managing the variables that can control product quality. Finally, the feedback from the selected customer site showed that the customer did witness improvements over time and that the manufacturing line's efforts were paying off.

SUMMARY

The original hypothesis of this study proposed that a manufacturing line could undertake a series of incremental improvements to its process that ultimately would result in improved product quality. It suggested that the impact of this continuous improvement activity could be witnessed through the gradual evolution of new tooling, the introduction of JITQC through employee involvement, and direct feedback from customer sites.

Once the hypothesis was defined, substantive data was gathered from a number of sources. This included an interactive manufacturing data base that collected relevant SPC information, a corporate materials tracking data base used to audit ship schedules and track defect reduction efforts, and information generated by an engineering task force seeking to standardize customer measurement capability. All three sources were tapped for useful information that could corroborate this study's hypothesis.

This study determined that all three categories did generate improvements that helped support the final goal of superior product quality. A continuous improvement project of this magnitude required a broad effort across many functional boundaries in order to allow the manufacturing line to develop a closer relationship to the customer that ultimately guided its future planning strategies. These multidisciplined efforts allowed the manufacturing line to become more proactive as an innovator for solutions to customer problems and to redefine its role as a valued corporate resource.

The manufacturing line is continuing efforts to follow up on all the improvements that impacted product quality. They are currently sponsoring efforts to formalize the organizational changes that will be needed to sponsor cross-functional work teams in manufacturing. This evolutionary trend will allow engineering to be used more as a resource and will encourage individual work teams to form to solve their own problems as they arise. Issues such as the amount of decision-making authority needed, the need for additional training, and how to best manage the performance objectives of individual groups while still encouraging local decision making are being discussed.

Additional efforts are still under way to encourage a proactive role between the manufacturing line's engineering staff and various customer sites. The engineering community is now actively participating in a number of experiments requested by customer sites that were designed to enhance product quality through value-added benefits. Customer sites are also beginning to utilize the manufacturing line as a resource to gain benefits through new experiments aimed at a six sigma quality program.

RECOMMENDATIONS

The information collected in the course of this study showed that the manufacturing line concentrated on areas that resulted in valuable improvements that went far beyond better product quality. American manufacturing efforts are approaching a crucial point in time, where substantial changes will be needed in the way companies manage their resources. Even a casual literature search today shows that one of the most neglected resources that a company can utilize in its arsenal of competitive strategies lies in the empowerment of its front-line work force.

The employee involvement survey and research readings all point to workers wanting a genuine participatory role in decision making which can improve the way things get done. Corporate strategies elicit lofty visions, but these visions will not turn into effective competitive weapons without the full cooperation of all employees.

In order to realize these benefits, the corporation must offer enhanced skills training and then encourage educated risk taking with decisions. People must believe that they have real control, or this will not promote innovative thinking.

Efforts have already started in selected areas to define what controls are needed to encourage this process. Management is currently encouraging the formation of self-sustaining teams, with members providing shared expertise and solutions to meet corporate goals and objectives. Management is also redefining its expectations and trying to measure progress through performance-related activities to link this undertaking to corporate strategy.

The movement to self-sustaining work teams follows as a natural evolutionary process, from a commitment to JITQC through employee involvement, but it has many potential pitfalls that need to be managed and explored for the best return. Another area offering many opportunities for improvement lies in creating a closer relationship with package vendors to better manage incoming package specifications.

In retrospect, the area of customer feedback was critical to the manufacturing line in this study, and efforts are under way to improve their incoming product measurements. This mutual exchange of information was one of the most beneficial aspects that returned valued to the corporation. The customer site benefited from the expertise provided from other areas within the corporation, and the manufacturing site developed the feedback necessary to focus on improvements most valued by the customer.

References
1. Zebrowski, Jim J. 1992. *Incremental Improvements in An Electronic Manufacturing Line and Its Impact On Product Quality: A Study of Improvements Introduced by New Tooling, JITQC through Employee Involvement and Customer Feedback.* Unpublished Master's Thesis, Lesley College, Cambridge, Mass.
2. Zebrowski, Jim J. February, 1993. SMT Product Quality Improvement: Methodologies and Data Analysis. Reed Exhibition Companies.

Appendix 10-A

Final Assembly Survey

Your help in answering the following questions will be useful in understanding change in final assembly. The survey will take about 10 minutes of your time to fill out, and all replies will be strictly confidential. Your participation in this survey is strictly voluntary, and the results from this study will be shared with you after the completion of the study.

Please place a check mark to the left of a category you choose in response to the following questions.

1. I have worked for Digital in the Final Assembly Area for:

    __ 0 to 6 months __ 6 months to 1 year __ Greater than 1 year to 2 years

    __ Greater than 2 years to 3 years __ Greater than 3 years

2. Area of responsibility:

    __ Line Technician __ ETG __ Supervisor __ Training __ Engineering

    __ Process Technician

Please place a check mark to the left of the word you select that best describes how you felt about each of the following statements as if you were asked about them last year. The five categories you can select from are as follows:

   __ Strongly Agree __ Agree __ Neutral __ Disagree __ Strongly Disagree

1. I am encouraged to help solve work-related problems.

   __ Strongly Agree __ Agree __ Neutral __ Disagree __ Strongly Disagree

2. People feel better about their jobs when asked to participate in solving job-related problems.

   __ Strongly Agree __ Agree __ Neutral __ Disagree __ Strongly Disagree

3. Engineering and ETG often involve line technicians in decision making that directly affects production.

   __ Strongly Agree __ Agree __ Neutral __ Disagree __ Strongly Disagree

4. We are informed about changes that affect the way we do our jobs.

   __ Strongly Agree __ Agree __ Neutral __ Disagree __ Strongly Disagree

5. We are encouraged to express our ideas/concerns about the way changes are introduced in manufacturing.

   __ Strongly Agree __ Agree __ Neutral __ Disagree __ Strongly Disagree

6. Engineering supports production to help solve out-of-control SPC issues.

   __ Strongly Agree __ Agree __ Neutral __ Disagree __ Strongly Disagree

7. I understand the value and importance of continuously improving the way work is done.

   __ Strongly Agree __ Agree __ Neutral __ Disagree __ Strongly Disagree

8. Communication on the manufacturing floor between Engineering, ETG, and Line Technicians is good.

   __ Strongly Agree __ Agree __ Neutral __ Disagree __ Strongly Disagree

9. Everyone on the manufacturing line works as a team to introduce new improvements.

   __ Strongly Agree __ Agree __ Neutral __ Disagree __ Strongly Disagree

The survey then repeated the same set of questions with the following opening explanatory statement.

Please answer the following questions in the same manner from the point of view of how you feel about these statements today.

It also finished the set of questions with a free-form question for everyone to answer in their own words.

Answer the following question in your own words.

The thing that would most help me improve my job performance would be:

_____

_____

_____

11

Improving Manufacturers' Distributor Performance Using QFD

Stephen M. Miu
Bull Worldwide Information Systems, Inc.

PROJECT BACKGROUND

As competition in the computer industry heats up, original equipment manufacturers (OEMs) are placed under increasing pressure to reduce inventories, shorten lead times, and reduce the supplier base by building partnerships with key suppliers to enhance teamwork. Likewise, the expectations of quality and service have been raised to new levels; service is the new buzzword and "competitive weapon" for the nineties.

Greene-Shaw Electronics has been servicing the area for many years, and has been a key supplier to Bull since the 1960s. However, business levels with Greene-Shaw were flat for a few years, due partly to the saturation of the computer market, but also due partially to the rising defect levels of inships received from Greene-Shaw. Measured in parts per million (PPM), these defects were manifested as administrative inaccuracies—primarily "wrong parts." These are mistakes that are easily corrected, though time consuming, expensive, and unnecessary. As a result, Bull purchasing agents (buyer-analysts) were not ordering new technology parts from Greene-Shaw and, in fact, considered targeting Greene-Shaw as part of the supplier reduction program.

From a business perspective, Material Quality Engineering (MQE) felt that eliminating Greene-Shaw as a supplier would be short-sighted—Greene-Shaw had been servicing Bull for over 30 years; the salespeople understood how Bull operated and typically provided good service for the buyer-analysts. Their proximity to both the Lawrence and Brighton manufacturing facilities made it an ideal partner for Bull's future growth. Similarly,

Greene-Shaw was the largest distributor of key components used on Zenith Data Systems (ZDS) products.

Hence, this project's goal was to reduce Greene-Shaw's 1990 PPM level to Bull by 40% in 1 year. This would be achieved by beginning a *total quality management* (TQM) program, specifically incorporating *quality function deployment* (QFD) to determine departmental efficiencies and ability to synergize. Timely deliveries of promised components, shipments that are correct and perceived to be the highest in quality, and sales and technical support before and after the transaction were desired improvements. As will be seen, this conservative figure was far exceeded.

IMPLEMENTATION

This project originally started as a joint movement to computerize all cross-reference lists from Bull part numbers to manufacturer commercial part numbers. This was due to the fact that there was an increasing number of administrative rejects from Greene-Shaw, despite a business level that was essentially flat. When this effort was completed with minimal impact to the PPM level, it became evident that the root cause needed to be addressed—more had to be done. It was at this time that MQE decided to approach both sides of management to propose and initiate a TQM/QFD program.

With management from both companies, MQE explained and set expectations of the project. As mentioned earlier, MQE chose QFD as the primary tool because, when employed in an environment that had little focus, QFD would prioritize the business features that needed attention and would provide clues for management's resolution. In this case, Greene-Shaw was to provide the energy and any funding that was to be required, while MQE set the direction and provided the expertise and education to meet the initial targets and encourage continuing feedback.

The improvement process was kicked off at Greene-Shaw Electronics with a general "all-hands meeting" that depicted the aim of the project, how we were to proceed, and what was expected of everyone. A movie, "Discovering the Future—The Power of Paradigms," was presented, and a pizza lunch was catered to kick off the "new mentality."

This initial process improvement team consisted of MQE, to function as facilitator and key process auditor; Greene-Shaw Electronics's VP of Sales and Marketing, to provide the constancy of purpose and management "buy-in" to get changes implemented; Greene-Shaw's outside sales manager and sales engineer, to gather feedback on the process from customers and inside staff members; and Greene-Shaw's management information system (MIS) engineer, to provide the data base and computer manipulations to track the progress of the company. The team was to meet informally on a biweekly basis at the Greene-Shaw location, and would last about 9 months. At that time, the seeds were to be planted, and the company would have the tools and skills needed to continue the process independent of MQE.

Both Bull and Greene-Shaw agreed that a complete analysis of the operation would be required to establish a baseline for improvement. Internal process audits were performed for each department: order placing, inside sales, inside sales execution, accounts receivable, warehousing, shipping, and (MIS). Included in this process were "dummy runs" through the process, monitoring how an order gets processed, and weekly meetings with personnel in each department, carefully scheduled to minimize impact on everyday business. In this forum, open-ended questions asking employees to list their responsibilities, internal and external customers, hindrances to productivity, and general improvement areas were reviewed, discussed, and formalized.

As a whole, the general consensus as far as barriers to daily responsibilities for the company as a whole included:

45% Need a "real-time" ability to talk to key people
35% Less paperwork
20% More legible paperwork
17% Traceability
13% Ability to pull sample requests quickly

Responses add up to greater than 100% due to multiple responses.

As far as the fuzzy definition of a "defect," we reached consensus; for ease in monitoring progress, a "defect" was anything that the customer wanted to return, regardless the reason or whose fault: incorrectly ordered, wrong part, overshipment, received early, and so on.

The general summary of responses solicited were easily predictable. Everyone had a small window of specialization, but "pretty much did a little of everything." Surprisingly, when the subject of customers arose, they all identified that their independent work impacted everyone within the organization.

My work impacts:

100% Everybody else at Greene-Shaw
 95% The customer

Responses add up to greater than 100% due to multiple responses.

This acknowledgment overcame a major hurdle, in that MQE and Greene-Shaw management were concerned that the workforce would be unmotivated and would not enthusiastically embrace the program.

ANALYSIS

As the project facilitator, MQE then created an Ishikawa (Cause and Effect) diagram, depicting "How Defects Occur" (Figure 11-1). Since distribution is a service operation, the "Four P's" were examined: Procedures, Policies, Plans, and People. In summary, consensus among the different departments was that management needed to address key items before any major improvement could take place: set expectations, establish policies and procedures to close the order administration process loop, and eliminate motivational shortcomings.

The preliminary analysis of internal concerns and MQE observations from the continuous audits were explained and discussed with all affected departments. At that time, recommendations were presented to management for action. These recommendations were to be the first of what would be an ongoing movement to refine the business. The preliminary recommendations were broken up as short-range and long-range plans. Short-range plans were changes to processes that could be implemented immediately, while long-range plans were changes that would require capital infusion and perhaps cause a temporary reduction in productivity.

Brainstorming meetings with the Bull buyer-analysts and Greene-Shaw's outside salespeople resulted in an agreement that, to increase business with both established and potential customers, salespeople must keep promises and take an earnest concern with the issues of the buyers. Storage and handling practices of "standard" product is also the key to shipping the right part to the right customer at the right time. This informal session was summarized in Figure 11-2, which segregates the qualities of distributorships that is important to each.

Using these results, the completed QFD matrix (Figure 11-3) determined our priorities. The QFD matrix is a visual aid that helps to benchmark one's products and services to those of competitors. The team's interpretation of the matrix shows that the administrative processes and delivery systems were most important in sending a quality product. In establishing an industrial position, the team determined that the term "quality" is very ambiguous, but, in discussion, it was generally agreed upon that there were the four key distributorships (in the area) that Greene-Shaw needed to target in order to increase business.

Examples of some short-range solutions were:

1. Establish new operating procedures.
2. Relabel shelves.
3. Standardize the order entry operation.
4. Acquire "small item" capital equipment.
5. Create a new quality reporting system on the computer.

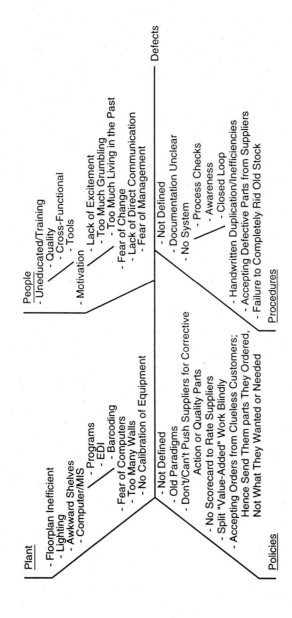

FIGURE 11-1. Ishikawa (C&E) diagram of defects.

Buyer:	1	2	3	4	5	6	7	8	10
Priority:									
1.	Delivery (part age)	Quality	History (response)	Delivery (part age)	Price	History	History delivery inventory Price	History (consistency)	Mfg. preference
2.	History (accuracy response tech support)	Relationship (tech support)		Price	History (flexible)	Price		Delivery (relationship)	History
3.	Quality	Delivery		History	Delivery	Delivery		Price	Price
4.		Price		Quality				Size	Delivery
5.									Quality
Preferred Distributors:		Time Heiland	Spartan Marshall Eagle	Sager Time Bell	GS	GS AIP GRubber	GS	Bibco Capstone Heiland Bell	
Comments:	*	a	b	c	d*	e*	*	fgh*	

KEYS TO COMMENTS

a ^ Standard products only vary about 5% in price
b ^ Spartan because local to Lawrence
c ^ Quality is last because haven't seen many problems, only administrative
d ^ GS lead times or corrective actions not timely because "we are at the mercy of the manufacturer"
e ^ GS jumps–response very good
f ^ No favorites—whomever has the parts, and can deliver it correctly
g ^ If a toss up who to give business to, order based on personal relationship with sales person
h ^ Price is a consideration only if there is a large price differential
* Signifies that "History" includes experienced parts quality

NOTES

- The buyer-analysts depicted here were solicited for input as to any criteria that they base distributor orders with.
- Only buyer-analyst #9 did not provide any feedback for this study.
- All buyer-analysts who responded made reference to the fact that purchasing from distributors is a very "intangible" ordeal.
- No one had set guidelines to follow, but overwhelmingly, the same key fundamentals were stressed: delivery, quality, history, and price. This fact was also reflected in the QFD Matrix.
- "Preferred Distributors" are dependent on the commodities that each buyer-analyst is responsible for.

FIGURE 11-2. Procurement criteria of purchasing agents.

Quality Characteristics

Customer Remarks	History	Company Size	Parts Quality	Delivery Systems	Admin Systems/Proc.	Rating of Importance	Company Now	Competitor x	Competitor y	Competitor z	Plan	Ratio of Improvement	Sales Point	Absolute Wt.	Demanded Wt.
						A	N				P		B	C	D
Ease of Business	⊙ 1.62	⊙ 1.62	△ .18	○ .54	⊙ 1.62	3	5	4	4	4	5	1.0	1.5	4.5	.18
On Time	△ .30	△ .30	△ .30	⊙ 2.7	⊙ 2.7	5	4	5	5	4	5	1.25	1.2	7.5	.30
Correct Parts	△ .39	△ .39	⊙ 3.51	⊙ 3.51	⊙ 3.51	5	3	5	5	5	5	1.67	1.2	10.0	.39
Technical Support	○ .27	○ .27	○ .81	△ .09	△ .09	2	4	4	4	4	4	1.0	1.2	2.4	.09
Low Price	△ .12	△ .12	○ .12	△ .04	△ .04	1	4	3	4	3	4	1.0	1.0	1.0	.04
Total	2.7	2.7	4.92	6.88	7.96	25.16							Total	25.4	100%
%	11	11	21	27	32	100%									
Company now															
Competitor x															
Competitor y															
Plan															

Main Correlations
⊙ 9 = Strong Correlation
○ 3 = Some Correlation
△ 1 = Possible Correlation

D = A × B × C
D = P
B ≠ P/N

FIGURE 11-3. QFD matrix.

For the inside salespeople, the order entry process now included a standard operating procedure (SOP) checklist to be conformed to by Bull buyer-analysts and Greene-Shaw's inside salespeople. The checklist, seemingly simplistic, was surprisingly effective—not from an informational context, but from an interpretational aspect, as information could be quickly assimilated by scanning predetermined zones on the paper (Figure 11-4). The short-range changes were implemented and adhered to, yielding immediate results.

One major shortcoming of the administrative processes was the duplication of paperwork. Salespeople worked with too many sheets of paper that conveyed the same information. All sheets were handwritten, hand copied, and passed on by hand. Lack of automation aside, there was no accountability, minimal attempts to clarify poor handwriting, and no prevention (which was partly addressed by the checklist). Sheets were lost or misplaced and were difficult to sort. Customer samples were errantly picked, which violated the integrity of the inventory management system, and, consequently, were not followed up by anyone. Orders were shipped without an institutionalized routine, which made tracking difficult.

The true nature of the project's difficulty was seen when recommended long-range process changes were presented and met with speculation or general reluctance from the management team:

1. Specialization of departmental duties.
2. Reorganizing of the functional floorplan.

The general patronizing, "Oh, is that right?" attitude was very noticeable. It became more and more evident that MQE's position of accountability without true authority was not desirable. This was probably due to the fact that, over time, management's lack of understanding and entrenched paradigms prevented effective management leadership of QFD. QFD/continuous improvement results were comparatively slow, frustrated enthusiasm, and eventually placed lower on departmental objectives. This stymied momentum.

PRIORITIZATION AND INSTITUTIONALIZATION

Fortunately, at Greene-Shaw Electronics, the team had fostered relationships throughout the organization that helped to facilitate and guide much of the

Improving Manufacturers' Distributor Performance Using QFD 197

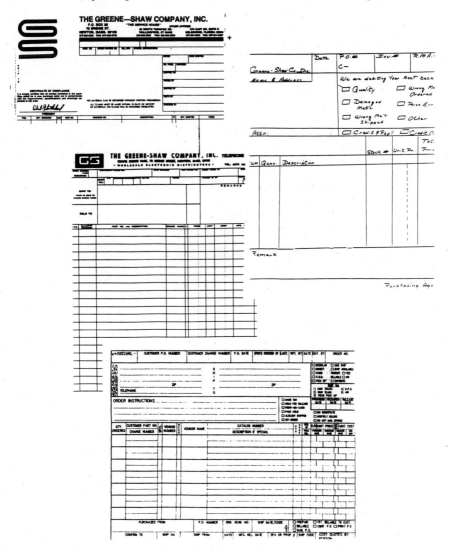

FIGURE 11-4. Sample paperwork, old and new.

push to improve quality. Together, it reaffirmed "quality awareness" in service again. This provided the necessary sense of urgency and constancy of purpose that management had to embrace if this project was to ultimately succeed.

As mentioned previously, the appointment of the VP of Sales and Marketing, responsible for all quality issues, was instrumental to the success of the project. He undertook several action items to compress the timeframe of our long-range solutions. Likewise, the manager of Outside Sales and outside sales engineers were instrumental in keeping the enthusiasm within the organization, while their rapid action and attention to detail minimized the impact of the defects that did get through to Bull.

Also beneficial were the increasing number of other customers' inquiries about Greene-Shaw Electronics's ability to support just-in-time (JIT), electronic data interchange (EDI), and even ISO 9000—all key components in today's corporate quality systems.

To remain competitive, the team emphasized that Greene-Shaw Electronics needed to take the threat to marketplace seriously. As the plan dictated, shortcomings were identified and attacked. In general, they had to identify why they were not meeting customers' present expectations and correctly anticipating future expectations and industry standards. What good points from their competitors can they emulate? How can Greene-Shaw reduce the returns rate, and how they can reduce overhead?

Among the many policy and process "tool" changes implemented were the following:

1. The creation of a complementary software-based quality tracking program, which provided accountability of order processing to the individual in each department. This new software enhanced the previously established computer program that summarized the dollar volume and shipment data that was being used as a tool for MQE.
2. A new order entry support program was implemented to electronically cross-reference part number information from customer to manufacturers' commercial numbers and revisions. Used in conjunction with their existing MIS, which tracks inventory, shipping and build requirements, and finance, Greene-Shaw Electronics is much better prepared to anticipate customer demand for components.
3. All old paper purchase specs and cross-reference lists have been isolated and put in storage, pending disposal.
4. EDI capabilities were expanded when their computer system was upgraded. Presently, they have seven communication lines dedicated to EDI applications.
5. Likewise, barcoding of inventory will be available to both Greene-Shaw Electronics and customers on a wide scale at that time.
6. Simultaneously, computer programs have been streamlined and made easier for everyone to access. The MIS department is now free from keypunching and data entry and is dedicated to computing and servicing the employees and EDI partners.

7. Documentation of job descriptions by the employees was initiated. Any gaps of responsibilities were then connected and assigned by management.
8. A formal organization chart was updated and circulated. This allows resources to be allocated to areas that require more attention.
9. Regular "quality" meetings will take place bimonthly to review successes and concerns of the company.
10. SOPs, which now include the follow-up calls on customers who requested information, have been formalized.
11. The office floor plan was rearranged. Barriers to communication, such as high partitions and glass walls, were removed. Sales support functions—purchasing, customer service, and billing—were then moved to be centrally located. Fax machines were also moved for easy access by all departments. Regional warehouse inventory was consolidated and relocated to the headquarters facility. Previously sporadic and small storage areas were moved into one area that is well lit and easily accessible. This undertaking was done to keep tighter control over quality (and processes in anticipation for ISO 9000 certification activity), consolidate inventory, and take advantage of volume component discounts. Obsolete and damaged components were purged.
12. Departmental sales and quality accomplishments are prominently displayed on the walls.
13. Four manual entry forms were consolidated into one. Worksheets are now kept by each salesperson in a queue to allow easier follow-up with customers.
14. Inside salespeople call a minimum of three customers a day to follow up inquiries and solicit feedback firsthand from the field. In this way, Greene-Shaw can proact and ensure customer satisfaction, instead of juggling with crisis management. This strategy has already identified incidences where customers were unhappy with the goods or services they had received and allowed its correction. Recognizing the customer this way has allowed Greene-Shaw Electronics to dispel concerns and enunciate their "customer service" dedication.
15. The Value-Added/JLC Cable Manufacturing subsidiary was moved into the Newton facility's former warehouse space, which proved to be more than adequate.
16. With the consolidation and change in ownership, the previous inefficiencies became very evident, as people found themselves underutilized. A reduction of staff cut costs, and systems homogenization with the new parent controlled materials flow even better.

Old and new floorplans are provided in Figure 11-5.

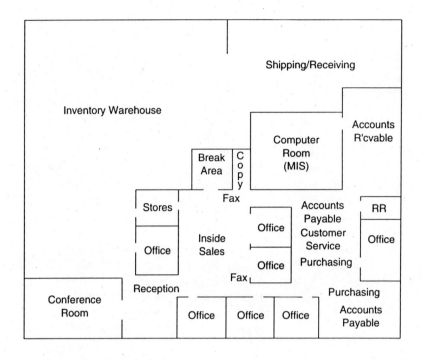
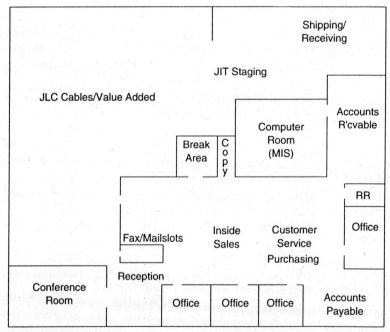

FIGURE 11-5. Floorplans: old (top) and new (bottom).

200

MONITORING/FEEDBACK

Of the approximately 50 distributors and representative companies that Bull deals with on a regular basis, Greene-Shaw Electronics ranked very favorably in 1991. At Bull, material quality and procurement performance is measured by defect PPM (i.e., how many pieces are defective when extrapolated to a million). Every month, the MQE supplier data base is queried and then the quality history is extracted. Similarly, the overall shipments information is printed out by Greene-Shaw's MIS, where the same calculations are performed. This would provide the team with information to track the effects of the TQM/QFD project with their overall customer base.

After 1 year, the overall trend of rejects to Bull was sharply downward (Figure 11-6), while the trend for Greene-Shaw business was downward overall as well (Figure 11-7). Though not as pronounced, the GS overall PPM decline was favorable.

These efforts to improve service quality have raised workforce morale to its highest level since the project's inception. The employees now feel that they have true accountability and impact on the company in setting its direction. This will undoubtedly improve Greene-Shaw's competitiveness and will manifest itself to be one of the many benefits to Bull. As presented

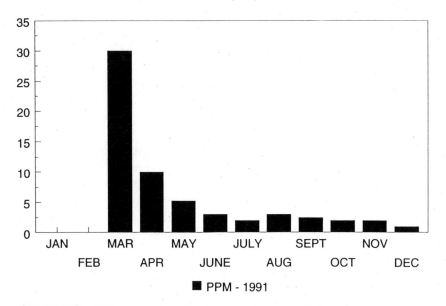

FIGURE 11-6. 1991 Greene-Shaw Electronics performance to bull.

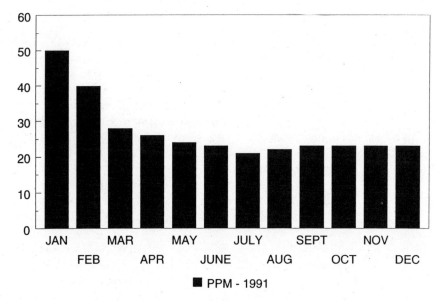

FIGURE 11-7. 1991 Greene-Shaw Electronics performance overall.

previously, both companies have already started to profit from the numerous process and attitude changes.

Annual "spikes" associated with rejected material due to regular inventory adjustments has been virtually eliminated. It is anticipated that, as the new process continues to monitor itself, any large fluctuations would signify a truly unusual event.

It is also expected that this overall downward trend will continue throughout the year and ensure that Greene-Shaw Electronics is one of the leading quality suppliers to Bull. After the new processes become institutionalized and have demonstrated effectiveness, MQE hopes to certify Greene-Shaw Electronics to the requirements and expectations of the Supplier Quality Assurance Program (SQAP), to become the only certified distributor for small passives and electro-mechanical devices to Bull's U.S. Manufacturing Division. This would further improve the production of ZDS and GCOS products at both Lawrence and Brighton manufacturing facilities, while maintaining high levels of quality products, service, and customer response.

Simultaneously, to improve the overall numbers, sales engineers will work to review all customer requirements and try to better understand their needs. For better understanding of quality processes, Greene-Shaw Electronics management has created a quality improvement team and has committed key personnel to participate at Crosby College. Ultimately, management's goal is to get ISO9000 certification and to set processes and documentation to the standards and expectations of the Malcolm Baldrige Award criteria.

These efforts to total quality and continuous improvement will benefit both companies in the long run and will undoubtedly improve our reestablished success partnership.

CONCLUSION

In today's economic uncertainty, OEMs need to be increasingly competitive in order to survive. This is especially true for a global company such as Bull. A firm supplier base with which it can rely on to provide products in a timely fashion and that has a contingency to deal with discrepancies immediately is essential to the rapid turnaround required for building printed wire assemblies.

Hence, true relationships must be manifested in order to maximize synergies. As such, much time was spent auditing Greene-Shaw, rebuilding and nurturing a partnership that will benefit both companies. Regular meetings take place to review and correct past problems and to anticipate future areas for improvement. Key lessons learned from this project were:

1. Maintaining employee motivation and involvement.
2. Continuing education for the workforce.
3. Correctly anticipate and interpret customer expectations and problems.
4. Emphasize prevention.

These are required in order for any continuous improvement program to succeed.

The overall costs incurred in setting up this program was minimal. Greene-Shaw Electronics purchased a new video monitor and contracted the removal and upgrade of certain walls and partitions that hindered communication and information flow. The building owner, as part of a predetermined renovation, installed new windows and lighting fixtures as part of the building's lease agreement. Some furniture rearrangement and cosmetic work was performed by the team on weekends.

The timing of the project was ideal. Aside from the potential loss of business from Bull, other key customers began to inquire about Greene-Shaw's implementation of quality processes and technology tools. This showed management that the demands from Bull were not out of line with the industry and that quality has become an underlying business philosophy for the 1990s.

The payoff of time and energy on this project cannot be measured accurately in a monetary manner, but rather in the efficiency of both companies when dealing with paperwork; the reduced handling of the parts and paper (via dock-to-stock and fewer Material Review Boards for discrepant material), accounting departments payable/receivable cycles (return materials authorizations), and so forth. These positive results enable MQE to expect good things from Greene-Shaw Electronics. Their open-mindedness and team-oriented thinking are all assets that any customer will look favorably upon. Assuming that they retain their strong conviction to quality and improvement, future relations will only improve.

References
1. Bill, N, and D. Lyman. 1991. QFD: a practical implementation. *Quality*, May 1991, pp. 36–40.
2. D'Alessandro, J. 1992. Leading by example: how EDI came to AVX. *EBN Purchasing Issues*, February 3, 1992, pp. P4 and P12.
3. Gryna, F. 1991. The quality director of the '90s. *Quality Progress*, May 1991, pp. 51–54.
4. Hansen, K. 1992. Judgment day arrives for suppliers. *EBN Purchasing Issues*, February 3, 1992, p. P14.
5. Feigenbaum, A. 1990. Implementing quality is a tall task for U.S. managers. *Electronic Business*, October 15, 1990, p. 44.
6. Forger, G. 1991. A move to carousels boosts orderpicking efficiency 118%. *Modern Material Handling*, April 1991, pp. 60–61.
7. Garwood, D., and M. Bane. 1991. Shifting paradigms mean a shift in management. *Modern Materials Handling*, April 1991, p. 37.
8. Goddard, W. 1991. Work cells enhance JIT/MRP2. *Modern Material Handling*, March 1991, p. 41.
9. Hauser, J., and D. Clausing. 1988. The house of quality. *Harvard Business Review*, June 1988, pp. 63–73.
10. Lanza, J., and J. Carbone. 1991. Six top business tips from motor city. *Electronics Purchasing*, October 1991, pp. 26–29.
11. McGrath, M. 1990. How to build in quality when resources are limited. *Electronic Business*, October 15, 1990, p. 59.
12. Rubinstein, S. 1991. The evolution of U.S. quality systems. *Quality Progress*, May 1991, pp. 46–49.
13. Sullivan, L.P. 1986. Quality function deployment. *Quality Progress*, June 1986, pp. 39–52.
14. Vasilash, G. 1991. Teams are working hard at Chrysler. *Production*, October 1991, pp. 51–52.
15. Votapka, T. 1992. Distribs thank Santa. *Electronic Buyers' News*, January 20, 1992, pp. 1 and 8.
16. Avnet income. sales slide. *Electronic Buyers' News*, January 27, 1992, p. 48.
17. BELL GROUP OPERATING PROFIT UP 34%. SALES 15% IN 2D QUARTER. *Electronic News*, February 3, 1992, p. 23.

18. Electronic data interchange. *Modern Material Handling*, February 1991, pp. 94 and D21.
19. Errors cut with bar codes, handhelds. *Modern Material Handling*, April 1991, pp. 78 and D17.
20. Ford tracks trucks with RFID. *Modern Material Handling*, February 1991, pp. 94 and D17–D18.
21. How do you keep the music playing? *Quality*, May 1991, pp. 26–28.
22. Jaco, Coughlin eye Greene-Shaw buy. *Electronic News*, March 16, 1992, p. 22.
23. Labeling system reduces errors. *Modern Material Handling*, April 1991, pp. 78 and D19.
24. Over 300 ways to solve productivity problems in plants and warehouses. *Modern Material Handling*, December 1990.
25. The quality imperative. *Business Week*, October 28, 1991.
26. Concurrent engineering. *IEEE Spectrum*, October 1991.
27. Barker, J. 1989. *Discovering the Future—The Power of Paradigms*. Burnsville, Minn.: Charthouse Publications.
28. Boothroyd, G. and P. Dewhurst. 1989. *Product Design For Assembly*. Wakefield, RI: Boothroyd Dewhurst, Inc.
29. Crosby, P. 1979. *Quality is Free*, New York: McGraw-Hill.
30. Drucker, P. 1985. *Innovation and Entrepreneurship*. New York: Harper & Row.
31. Peters, T. 1982. *Thriving on Chaos*. New York: Harper Collins Publishers.
32. Shina, S. 1991. *Concurrent Engineering and Design For Manufacture of Electronic Products*. New York: Van Nostrand Reinhold.

12

Measurement System Analysis

C. F. Chrisafides
Alternate Circuit Technology, Inc.

There are many places where the word "quality" is heard today. If one were to watch television for an hour, he or she would most likely hear the commercials say the word "quality." If someone opened up a phone book to the yellow pages, he or she would probably see the word "quality" in several of the advertisements. We live in a quality-minded society. But, believe it or not, it was not always like this.

Back in the 1940s, Dr. W. Edwards Deming and Dr. Joseph Juran both knew, preached, and practiced some new-fangled techniques for increasing and maintaining quality in manufacturing industries. The methodologies had actually been in existence for 20 years, since Walter Shewhart practiced them in the 1920s at Western Electric in Andover, Massachusetts. But back then the war was over and business was booming. American industry would typically make ten parts to yield five. It didn't matter. The cost of quality could not be justified.

Meanwhile, Japan was in a crisis state. Troubled with a rock-bottom post-war economy and with most company CEOs behind bars for war crimes, middle managers were desperate for ways to help them get back on their feet—effectively and efficiently! At that time in history, General McArthur was occupying Japan. Having known of Deming as a prominent statistician and census organizer, he initiated a team, including Deming, Juran, and several others, to try and help Japanese industry get back on its feet.

The Japanese listened with open ears and open eyes. They took these tools to heart, as they listened and learned and became effective. Their employees became empowered, trained in *statistical process control* (SPC). They initiated teams whose agendas were towards common goals. They grabbed hold of the concept of Total Quality in the 1950s, when Dr. Armond Flagenbaum introduced the idea. In 1962, a quality consultant for Toyota, Dr. Kaoru Ishikawa, expanded on Deming's and Juran's ideas, making Toyota the company that it is today.

Currently, the choice of what kind of electronic equipment or automobile to buy, based on performance and reputation, is won by the Japanese. Is it a coincidence that, while Americans were too busy to listen to ways for improving quality (because the baby-boomers had money to spend), Japan became such an economic giant and is now trying to buy up America and its businesses? Who would have known?

Now everyone knows. The "cat's out of the bag." America has come to realize the need for continuous quality improvement. We have become a quality-minded culture. Thus, the reference to quality is seen in everyday life—advertisements, television, and so on.

But what makes up this quality, this thirst for continuous improvement? What ties it all in together? It is made up of all of the tools, techniques, methodologies, and experiences of the great visionaries previously mentioned, as well as the trials and tribulations of their fortunate apprentices.

Of these tools and techniques, *total quality management* (TQM) is the foundation of the house to be built. TQM, the "can opener" as some people call it, is an acronym that stands for none other than, in the author's eyes, common sense. Incorporating teams and teamwork among all levels of a workforce, nurturing cohesiveness and synergism in solving problems, breaking down every simple and complex process into its smallest elements to search for points of intervention and gates, benchmarking and showing marked improvements that become documented and standardized as the building blocks of the foundation—these are the premises on which Deming and Juran based their visions. But remove the laced frame away from this painted picture, for these things cannot happen without commitment and effort. Where the old school says, "If it ain't broke don't fix it," the new school teaches, "To stand still is to go backwards." So work is a big element. The term "can opener" refers to the opening of cans of worms. However, if these cans are not searched out, opened, and spread out for all to see, they cannot be fixed. This is TQM—the stage-setter for continuous quality improvements, the basics of working smarter instead of harder.

There are several other sets of tools. There is *statistical quality control*, which effectively provides feedback on process and products and monitors the behavior of the manufacturing process as a whole to enable the prevention of problems rather than the off-line detection of problems. Also important is, *design of experiments* (DOE), which enables the optimization of a process in an efficient and timely manner. The list goes on and on. But one question comes to mind here: Of all of these tools, what ways and means exist by which these tools are based? In other words, how does one qualify his or her system of qualification? Through *measurement system analysis.*

Measurement system analysis is the common ground upon which all improvement and qualification techniques lay (see Figures 12-1 and 12-2). It is the media for which all decisions must be based. It is being able to confidently quantity a benchmark, a success, or a failure. Measurement system analysis becomes the mold that controls quality as it is perceived, thus making it vital to any improvement.

There are many tools that fall under the TQM umbrella, which, when used correctly and for the right reasons, can improve quality tremendously. But, whether the quality tool chosen is SPC, DOE, or just plain old inspection through *approved quality levels* (AQL), results should always be qualified through a measurement system. But, for all of the thought and work that goes into these wonderful techniques and the additional cost of quality for these so called "non-value-added" steps, has anyone stopped to analyze the capability of the measurement system itself? How accurate is it? How is its precision as compared to the tolerance of the parameter being measured? Is the calibration frequency sufficient (or overkill)? What

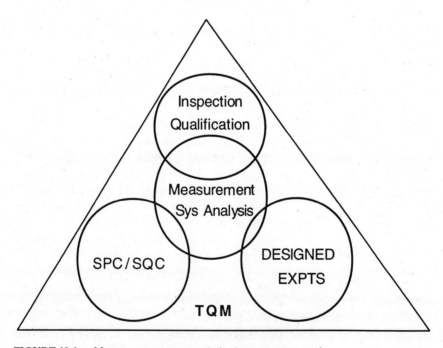

FIGURE 12-1.   Measurement system analysis: the common ground.

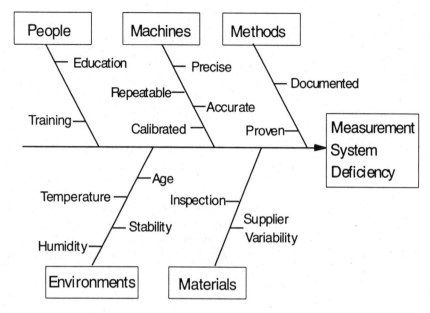

FIGURE 12-2. Cause and effect diagram for measurement system deficiencies.

about operator variation? Is training an issue? This chapter will attempt to answer these and other questions about a very important issue: *measurement system analysis*.

MEASUREMENT SYSTEM ANALYSIS

Measurement system analysis is a fairly simple combination of experience, observations, and statistical techniques that can be used in conjunction or separately to qualify any system used to make decisions based on results. A measurement system can be defined as a system consisting of an operator and/or a gage, which in conjunction can quantify a resultant effect.

A measurement system can be a simple ruler used by a technician to measure the length of "widgets as they come off a production line." A more complex measurement system may be using a hand-held 60x microscope to measure circuit trace width after etching it into a rigid circuit board, or even using X-ray fluorescence to measure metallization thickness of gold after plating jewelry, to make sure the outcome is within specifications. These are all treated basically the same way when evaluating measurement systems.

Measurement system analysis can be broken down into the following categories:

1. Accuracy.
2. Discrimination.
3. Repeatability.
4. Reproducibility.
5. Gage R & R.
6. Linearity.
7. Stability.

Each of these topics plays a distinct role in evaluating the measurement system—for example, when a measurement system is not properly calibrated (accuracy), when deterioration to a measurement system exists due to wear or environmental conditions (stability), when a training issue exists due to different people using measurement equipment in different ways (reproducibility), or when the ability to measure small differences (reliability) is jeopardized due to the scale of the measurement system (discrimination). All of these factors must be evaluated when qualifying a measurement system (see Figures 12-1 and 12-2).

ACCURACY AND DISCRIMINATION

Accuracy

Measurement systems are usually calibrated to a known standard. For example, when calibrating a micrometer using a known standard of thickness, one would perform the measurement and adjust the micrometer to read the known value, if it was in fact reading some other value. If one were to take a series of measurements on this standard and calculate the mean (X) value, this would be known as the "accuracy" of the micrometer.

A common definition of accuracy is the ability of any instrument to conform to the true value of a given phenomenon, as measured by the deviation of the average observed value from the accepted reference or standard.

Note: Accuracy should not be confused with the term "inaccuracy," which refers to the cumulative error, which is related to drift, nonlinearity, repeatability, stability, environmental effects, and so on.

One may purchase a multi-spindle CNC drill machine that is specified as having an accuracy in the x and y planes of .0001 (inch per inch). The manufacturer calculated that number by simply drilling a series of holes (perhaps 100) that are programmed to known locations. These holes are then measured. The average of the deviation away from the known programmed locations of the holes is the accuracy of the drill machine.

Accuracy is an excellent candidate for monitoring through SPC by use of control charts. Of course, this will depend on how critical the process is. For example, a TAB surface mount machine must be extremely "accurate" for automated circuit board population; thus, this would be a good candidate.

Discrimination

What happens when the measurement system being used is not precise enough to reflect variation in the parts being measured? For example, if a widget has specifications for its length of .005 inches +/−.0005 inches (tolerance spread of .0045 inches −.0055 inches) and the measurement system consists of a ruler that can measure accurately in increments of .01 inches, is it possible to pick up variation in the parts? Absolutely not! This is an exaggerated example of problems with measurement discrimination.

"Measurement discrimination is known as the technological capability of a measurement system to adequately differentiate between values of a measured variable."

Inadequate discrimination within a measurement system is overlooked more times than not. This is one of the most critical factors of measurement systems. If specifications for a parameter or characteristic are very fine, they are usually critical or else they would not be so tight. A good rule of thumb would be to have ten increments, inclusively, from the lower spec to the upper spec (see Figure 12-3). Case study A (in "Case Studies" section of this

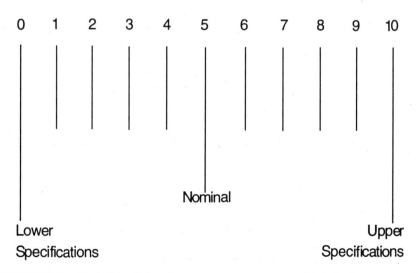

FIGURE 12-3. Rule of thumb for adequate measurement system discrimination.

chapter) illustrates an example of measurement system discrimination inadequacy.

REPEATABILITY

Variation within the equipment used to perform measurements can often contribute greatly to problems associated with the measurement system. If one operator were to measure ten parts three times and record them, he or she could calculate any variation present in the equipment (see Table 12-2) Assuming that the operator has been performing this measurement task for a discernible amount of time, the ranges in the measurements of the ten independent parts (including the average range) shall reflect variation within the measurement equipment. Performing this same test with several operators and comparing the average ranges will provide confidence in this measurement system evaluation.

This variation that is attributed to the measurement equipment is known as repeatability, or equipment variation (EV). This is defined as 5.15 sigma (5.15 standard deviations) range, which basically estimates the spread that covers 99% of the measurement variation due only to the measurement equipment. The ranges from the measurements on the same sample by the same operator are used to ascertain the standard deviation for repeatability (σ_{EV}). From statistical quality control, recall that ranges and standard deviations are related by the formula

$$\sigma_{EV} = \frac{\bar{R}}{d_2}$$

Here d_2 is a statistically derived constant (given) which is dependent on the sample size (number of trials) used to calculate a single range and $\bar{R}$ is the average range. Thus, repeatability is defined here as 5.15 σ_{EV}, or the "99% spread."

$$\text{Repeatability} = 5.15\ \sigma_{EV} = 5.15 \left(\frac{\bar{R}}{d_2}\right)$$

d_2 values are given in Table 12-1.

Derivation of Repeatability, or Equipment Variation (EV)

On the R&R report (Figure 12-4), the range for the trials of each operator are calculated. Then the average range is calculated for each operator ($\bar{R}, \bar{R}_2, \bar{R}_3$, etc). Finally, the average of the averages of the ranges, or $\bar{\bar{R}}$, is calculated, as shown in Table 12-2.

TABLE 12-1.   First 10 Values of d_2 and d_2^*

n	d_2	d_2^*
2	1.128	1.410
3	1.693	1.906
4	2.059	2.237
5	2.326	2.477
6	2.534	2.669
7	2.704	2.827
8	2.847	2.961
9	2.970	3.076
10	3.078	3.178

These measurement ranges on the same sample performed by the same operator are again used to figure out the standard deviation for repeatability (σ_{EV}). Therefore, σ_{EV} is calculated by using the equation

$$\sigma_{EV} = \frac{\bar{\bar{R}}}{d_2}$$

What Are K Factors and How Do They Effect R&R Studies?

We now define the following:

$$\text{Repeatability} = 5.15\,\sigma_{EV} = 5.15\left(\frac{\bar{\bar{R}}}{d_2}\right) = K_1\bar{\bar{R}}, \text{ where } K_1 = \frac{5.15}{d_2}$$

TABLE 12-2.   Repeatability

	Operator A			
Part No.	Reading One	Reading Two	Reading Three	Range
Part 1	85.3	85.8	85.5	0.500
Part 2	86.3	88.6	86.3	2.300
Part 3	88.8	89.0	88.8	0.200
Part 4	84.8	87.8	87.5	3.000
Part 5	88.2	88.5	88.3	0.300
Part 6	86.1	84.8	85.1	2.100
Part 7	86.8	87.4	87.4	0.600
Part 8	88.2	88.7	88.4	0.500
Part 9	87.0	87.2	87.2	0.200
Part 10	86.1	86.2	86.2	0.100
				$\bar{\bar{R}} = 0.980$

TABLE 12-3. K_1 Factors When (# Operators) $\times$ (# Samples) $\leq$ 15

No. of Trials	(# Operators) $\times$ (# Samples)													
	3	4	5	6	7	8	9	10	11	12	13	14	15	16
2	4.19	4.26	4.33	4.36	4.40	4.40	4.44	4.44	4.44	4.48	4.48	4.48	4.48	4.50
3	2.91	2.91	2.96	2.98	2.98	2.99	2.99	2.99	3.01	3.01	3.01	3.01	3.01	3.05
4	2.43	2.44	2.45	2.46	2.46	2.48	2.48	2.48	2.48	2.49	2.49	2.49	2.19	2.50
5	2.16	2.17	2.18	2.19	2.19	2.19	2.20	2.20	2.20	2.20	2.20	2.20	2.20	2.21
6	2.00	2.00	2.01	2.01	2.02	2.02	2.02	2.02	2.02	2.02	2.02	2.03	2.03	2.04

Notes:
1. If possible, select the number of operators and samples for (# operators) $\times$ (# samples) to exceed 15.
2. If (# operators) $\times$ (# samples) $\leq$ 15, enough trials should be run to avoid error. Otherwise, the estimates may be imprecise.
3. This form is a modification of the General Motors long form.

For example, in a study with three trials (as seen in Table 12-3), the sample size for each range is 3. The tabulated value of d_2 for a sample size of 3 is 1.69. Thus,

$$\text{Repeatability} = 5.15 \, \sigma_{EV} = 5.15 \left(\frac{\bar{\bar{R}}}{d_2} \right) = 5.15 \left(\frac{\bar{\bar{R}}}{1.69} \right) = 3.05 \, \bar{\bar{R}}$$

This is the source of the $K1$ factors shown in Table 12-3 for a given number of trials. However, Duncan (1974) has shown that this estimation procedure must be corrected when there are a small number of ranges (2, pp. 145 and 190). Since there is a range calculated for each operator-sample combination, the product of (# operators) $\times$ (# samples) is the number of ranges. If there are at least sixteen such ranges, the usual estimation procedure using d_2 is appropriate. In such a case, the values of K_1 from Table 12-3 are correct.

REPRODUCIBILITY

Reproducibility, or *appraiser variation* (AV), is assessed by observing the variations among the several operators measuring the same samples. This type of variation is usually composed of different techniques or styles that different people may have developed over time, and generally does not contribute greatly to measurement system inconsistencies. However, many times an analysis of the measurement system will show a large amount of variation due to reproducibility. In these cases, this is usually due to lack of training. This will be seen later in the chapter.

K_2 Factors

The same rationale applied to K_1 factors is also applied to obtain the K_2 factors on the R&R form. The K_2 factor is used to ascertain the standard deviation of the operators' averages, leading to the estimation of reproducibility. The range of the operators' averages is labeled "$\bar{\bar{X}}$ diff."

Again,

$$\sigma \text{ operator averages} = \frac{\bar{R}}{d_2} = \frac{\bar{\bar{X}}_{\text{diff}}}{d_2}$$

Such a small number of ranges (one) forces us to use d_2^* instead of d_2.
Then,

$$\sigma \text{ operators' averages} = \frac{\bar{\bar{X}}_{\text{diff}}}{d_2^*}$$

(d_2^* can be found in table 12-1.)

Reproducibility is defined as a 5.15 σ range. Then, using the same basis for the σ of operators' averages,

$$5.15 \; \sigma \text{ operators' averages} = \frac{5.15 \; \bar{\bar{X}}_{\text{diff}}}{d_2^*} = K_2 \bar{\bar{X}}_{\text{diff}}, \text{ where } K_2 = \frac{5.15}{d_2^*}$$

Other values for K_2 on the form are estimated similarly. This procedure provides degrees of freedom = (# operators − 1). Thus, using two operators, there is only one degree of freedom for the estimate. This is why it is recommended to use three or four operators if practical.

Note that, if there is only a single operator, it is not possible to calculate the K_2 factor, and thus there is no AV value.

Deviation of Reproducibility (AV) Calculation

An R&R study is a designed experiment in which the factors are 1) samples and 2) operators. The trials are nested within the operators and samples. It is assumed that:

1. The samples are random representatives from a very large population of samples.
2. The operators are random samples from a very large population of possible operators.
3. The interaction of operators and samples does not exist. The calculation of the upper control limit of the ranges on the R&R form is an attempt to confirm this assumption.

4. The study has been performed in such a manner, using random order of tests, that the pooled variation of the trials for each operator's sample combination is an estimate of the error mean square in an analysis of variance.

Then the ANOVA values, with expected mean square is:

Source	Expected Mean Squares
Samples	$\sigma_{EV}^2 + or\ \sigma_s^2$
Operators	$\sigma_{EV}^2 + rn\ \sigma_{AV}^2$
Error	σ_{EV}^2

Where n = number of samples
o = number of operators
r = number of trials

It is known that

$$\sigma_{EV}^2 = \left(\frac{\bar{R}}{d_2}\right)^2$$

The variance in operators' averages, however, is given by

$$\left(\frac{\bar{\bar{X}}_{diff}}{d_2^*}\right)^2$$

Since each operator's average is the average of $(r \times n)$ measurements, then using the ANOVA expected mean squares,

$$\sigma_{operators'\ averages}^2 = \left(\frac{\bar{\bar{X}}_{diff}}{d_2^*}\right)^2 = \frac{\sigma_{EV}^2}{rn} + rn\ \sigma_{AV}^2$$

Where the variance of an average of "m" observations is $\frac{\sigma^2}{m}$,

Thus, $\sigma_{operators'\ averages}^2 = \sigma_{AV}^2 = \left[\left(\frac{\bar{\bar{X}}_{diff}}{d_2^*}\right)^2 - \frac{\sigma_{EV}^2}{r \times n}\right]$

This is what is happening in the calculations when manually performing an R&R form. The only difference is that the form is working with 5.15 times the standard deviations, thus using K_1 and K_2. This results in

$$EV = 5.15\ \sigma_{EV} = K_1 \bar{R} = \text{equipment variation}$$

$$(AV)^2 = 5.15\ \sigma_{AV}^2 = 5.15 \left(\frac{\bar{\bar{X}}_{diff}}{d_2^*}\right)^2 - \frac{\sigma_{EV}^2}{rn} = \text{appraiser variation}$$

$$(EV)^2 = (5.15)^2 \sigma_{EV}^2$$

$$(AV)^2 = (K_2 \bar{\bar{X}}_{\text{diff}})^2 - \frac{(EV)^2}{rn}$$

$$\text{Then, } AV = \sqrt{K_2 \bar{\bar{X}}_{\text{diff}})^2 - \frac{(EV)^2}{rn}}$$

For discussion on the respective roles and advantages of analysis of variance versus $\bar{X}$ and R charts, refer to Duncan [2, pp. 683–684].

GR&R

Gage repeatability and reproducibility (GR&R) is the combination of the variation seen within the measurement equipment and the variation between the various appraisers using the equipment. This combinational variation is mathematically calculated and compared to the tolerance spread, or the area between the upper and lower specification limits of variables being measured and expressed by a percentage. Figure 12-4a illustrates a GR&R data collection and analysis worksheet.

There are several steps that must be taken when performing an R&R study:

1. Ensure that the gage, or measurement equipment, has been calibrated.
2. Choose at least two operators who normally perform the measurement system to be evaluated.
3. Choose samples to be measured in the analysis.
4. Have the first operator measure all of the samples once in a random order.
5. Have the second operator measure all of the samples once in a random order.
6. Continue until all of the operators participating in the study have performed "trial 1", measuring all of the samples once.
7. Repeat steps 3 through 6 for the number of trials chosen, ensuring that the results of the previous trials remain unrevealed to the operators.
8. Using the form shown in Figure 12-4a (or equivalent software), record all of the data and determine the statistics of the R&R study.
9. Evaluate the results.

The following are important factors to consider when planning, performing, and evaluating a GR&R study:

- When operators are performing sample measurements, it is critical that they replicate normal operating procedures (as they would normally do). Only then can a true indication of normal conditions be seen.

REPEATABILITY & REPRODUCIBILITY REPORT

Part No. & Name		Gage Name		Date	
Characteristic		Gage No.		Performed by	
Specification		Gage Type		Unit of Measure	

Oper.	A				B				C				D			
Sample	Trial 1	Trial 2	Trial 3	Range	Trial 1	Trial 2	Trial 3	Range	Trial 1	Trial 2	Trial 3	Range	Trial 1	Trial 2	Trial 3	Range
1																
2																
3																
4																
5																
6																
7																
8																
9																
10																
Totals																

SUM $\bar{X}_a$ $\bar{R}_a$ SUM $\bar{X}_b$ $\bar{R}_b$ SUM $\bar{X}_c$ $\bar{R}_c$ SUM $\bar{X}_d$ $\bar{R}_d$

$\bar{R}_a$	
$\bar{R}_b$	
$\bar{R}_c$	
$\bar{R}_d$	
SUM	
$\bar{\bar{R}}$	

MAX. $\bar{X}$	
MIN. $\bar{X}$	
$\bar{X}$ DIFF.	

D_4	NO. TRIALS
3.27	2
2.58	3
2.28	4

($\bar{\bar{R}}$) • (D_4) = UCL_R *

REPEATABILITY = EQUIP. VARIATION — % TOLERANCE ANALYSIS

EV = ( )( ) $\bar{\bar{R}}$ • K_1 σ_{EV} = EV / 5.15 = %EV = 100 [( )/( )] EV TOLERANCE

EV = %EV =

TRIALS	2	3	4
K_1	4.56	3.05	2.50

REPRODUCIBILITY = APPRAISER VARIATION *** — % TOLERANCE ANALYSIS

AV = ($\sqrt{\;}$ ( )^2 - [( )^2/(• ( )] X_{DIFF} K_2 EV n r • = %AV = 100 [( )/( )] AV TOLERANCE

σ = AV / 5.15 = %AV =

OPER.	2	3	4
K_2	3.65	2.70	2.30

REPEATABILITY & REPRODUCIBILITY (R&R)

R&R = $\sqrt{(\;)^2 + (\;)^2}$ EV AV = %R&R = 100 [( )/( )] R&R TOLERANCE

σ = R&R / 5.15 = %R&R =

*Limit of indiv. R's. Circle those that are beyond this limit. Identify the cause and correct. Either a)repeat these readings, using same oper. & unit or b)discard values and reaverage/recompute R & the limiting value UCL.

***If a neg. value is calculated under the sq. root sign or if there is only one operator, then AV = O

FIGURE 12-4a. Repeatability and reproducibility report.

Measurement System Analysis 219

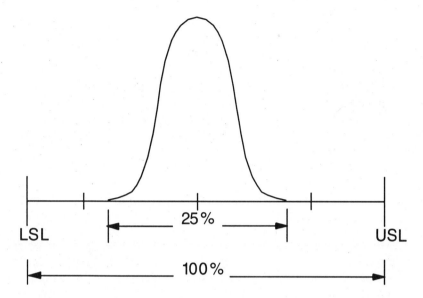

FIGURE 12-4b. This example depicts the distribution representing the combination of equipment and appraiser variation as it compares to the tolerance spread of the samples measured. The % R&R is 25%. An acceptable amount in this case.

- If calibration is performed at any set intervals (e.g., before every measurement), make sure that this is also done when evaluating any measurement system. Otherwise, unfair and/or unrepresentative bias may skew the data.
- A good rule of thumb for the number of samples to be measured is ten, and the number of operators used to perform the evaluation must be at least two. Also, just as in any statistical evaluation, the larger the amount of data (i.e., the more operators), the more confidence one can have in the overall evaluation.
- Samples to be used should, for best results, be of the same manufacturing lot (if possible) and should have the same upper and lower specifications (or at least the same tolerance spread).
- Once having calculated the *grand average range* ($\overline{R}$), or the average range of the separate average ranges (R) for each trial, it is important to calculate the upper control limit for the range (UCLR) and compare the individual ranges to it. Any of them that fall out of the UCLR must be given an assignable cause and be redone to ensure statistical control.

What is the outcome of the GR&R Study?

Once R&R is calculated, it is mathematically compared with the actual tolerance spread for the parts that were measured, and then given a percentage value. This percentage now indicates how much of the tolerance spread is absorbed by the variation within the measurement system (see Figure 12-4b).

Typically, an outcome of an R&R analysis of 30% or less is said to be acceptable. But this really depends on how critical and precise a measurement system is or should be. For example, visual inspection, such as reading a 60x hand-held scope, has a much greater potential for variation than, say, reading a ruler. Visual inspection using optical measuring equipment is known for having an average of about 24% R&R, an acceptable amount of variation. Again, this depends on the level of accuracy needed in the system.

When to Perform R&R Studies

There are typically two scenarios where R&R studies should be done. The first and most frequent is when a problem exists. When the outcome of a process proves inconsistent or is not meeting requirements, the first thing that should be done is to perform an analysis on how the results of the process are quantified. There are many cases where engineers and quality professionals spend countless hours and dollars on characterizing the process through SPC and optimizing the process outcomes through techniques such as design of experiments. These techniques are excellent tools; however, your outcome is only as good as the method you use to quantify it. Thus, the first step in nonconformance analysis should be to evaluate the measurement system.

Secondly, almost all measurement or processing equipment is calibrated on some type of a scheduled basis in order to ensure its accuracy. However, calibration does not account for stability or linearity (to be discussed in the next section) or for operator variation between cycles. The author recommends that GR&R studies be performed on the critical measurement systems on a planned and systematic basis on opposite schedules with calibration cycles. GR&R studies are very quick (especially with the many software packages available today) and painless, and they can provide a lot of worthwhile insight into critical measurement systems.

LINEARITY AND STABILITY

Measurement systems are very susceptible to other factors not yet mentioned here. These factors include environmental effects on the equipment and the deterioration that this can cause over time. Also, the accuracy of a measurement system over a range of values can vary. Does the measurement system lose accuracy over time? Is it more accurate at the lower or higher values?

Linearity

If a machine utilizes an oscilloscope as time domain reflectometer (TDR), its purpose is to measure the impedance of a circuit trace (in ohms). For example, suppose a GR&R study was done on ten samples of 40 ohms (+/−5%) and the %R&R was 5%. From this information it would be deduced that this measurement system has exhibited very little variation and is therefore acceptable. Now, if the same operators used in that first R&R study were to perform the study three more times on 60-ohm, 80-ohm, and 100-ohm samples and got an outcome %R&R of 10%, 20%, and 40%, respectively, how would this increase in variation be explained?

When the range of sample values is large, typically in comparison with the individual tolerance of the samples, linearity is something that should definitely be considered. In most cases, a given piece of measurement equipment is specified to have a certain amount of accuracy. This is usually either the average over a range of values or a tolerance about a typical "nominal" value. A range is never given! This is not an uncommon scenario and should be considered if a measurement system deals with a large range of values.

Stability

Temperature, humidity, wear, and other factors can also greatly affect any measurement system. Just like any other process being characterized and/or monitored by using statistical techniques, a critical measurement system can and, in some cases, should be monitored by use of a simple control chart (either X and R or Individual Measurements with Moving Range). Once the distribution for the measurement system is approximated, using six standard deviations about the mean (plus and minus 3), points falling outside of these calculated control limits can be designated as assignable causes so that they may be isolated and identified, qualified, and (hopefully) eliminated.

CASE STUDIES

The following case studies were performed by the author, a Quality Control supervisor at Alternate Circuit Technology, Inc. (Ward Hill, Massachusetts).

Alternate Circuit Technology (ACT) is a designer and manufacturer of prototype and preproduction complex leading-edge multilayer print wiring boards. The main facility operates out of 20,000 sq. feet, with a staff of 270. ACT is an industry leader, with a strong commitment to quality. Their success over the years has been the result of researching the needs of a competitive future. This is why each employee has been trained and certified through an intensive 21-hour SPC course, and TQM training has been given to every supervisor, manager, engineer, and company representative. Management has empowered employees at every level to organize task teams in each area, for each shift. These teams work toward problem isolation and rectification, for continuous quality improvement. Management philosophy is based on teamwork, and this makes synergism successful at all levels.

There are two case studies discussed here that illustrate real-time examples of some of the preceding topics. They are both taken from the printed circuit board manufacturing industry, where the author has had most of his experience. There are many other case studies on file, and the author encourages anyone to feel free to contact him, so that they may learn from his mistakes, thus making their own efforts more focused.

Case Study 1

This first case study deals with trying to optimize a process of etching circuit traces onto a copper-clad sheet of laminate using a photo-imaged circuit pattern as an etchant-resist. The requirements of this process in this situation were to achieve accurate line width and spacing requirements. Here, the technology was .005-inch lines spaced .005 inches apart (termed "5 and 5" technology in the circuit board manufacturing industry).

When the process began yielding inconsistencies in line loss, a team was formed to evaluate and analyze the process for improvement. Too much line loss yields "opens" in the circuit traces, while too little line causes "shorts." Where these circuit patterns were extremely dense, one open or short condition usually resulted in a scrap condition at an electrical test, performed after several of these layers become bonded together and processed.

The team brainstormed to form a cause-and-effect diagram in order to isolate all potential causes for inconsistent line loss, the foremost being the measurement system. Upon evaluation of the measurement system, it was realized that it had several shortcomings, the biggest being its discrimination.

A 60x hand-held scope with increments of .001 inches was being used to measure line loss. The specification limits for the parts that were being measured was .0045–.0055 inches. This translates into a reading that is either good or bad: no in-between! as stated previously, a good rule of thumb is to have the ability to discriminate with ten increments, from the lower specification limit to the upper specification limit. The operators were trying to extrapolate to a measurement of .00025 inches, using best judgement, which was essentially allowing bad product to be acceptable and vice versa.

100x scopes with reticles reading in increments of .0001 inches allowed the operators to confidently discriminate between measurements. This also allowed for SPC to be performed with confidence on line loss, so that assignable causes for unnatural variation of line loss could be isolated and eliminated.

TABLE 12-4.

	Operator 1			
Sample no.	Trial 1	Trial 2	Trial 3	Range
1	89.78	89.85	89.95	0.170000
2	90.68	90.63	90.74	0.110000
3	93.59	93.70	93.71	0.120000
4	91.66	91.84	91.91	0.250000
5	92.93	92.96	93.04	0.110000
6	89.04	89.04	89.20	0.160000
7	91.74	91.88	91.90	0.160000
8	93.11	93.13	93.19	0.080000
9	91.65	91.62	91.72	0.100000
10	90.41	90.49	90.43	0.080000

	Operator 2			
Sample no.	Trial 1	Trial 2	Trial 3	Range
1	89.78	89.85	89.95	0.170000
2	90.68	90.63	90.74	0.110000
3	93.59	93.70	93.71	0.120000
4	91.66	91.84	91.91	0.250000
5	92.93	92.96	93.04	0.110000
6	89.04	89.04	89.20	0.160000
7	91.74	91.88	91.90	0.160000
8	93.11	93.13	93.19	0.080000
9	91.65	91.62	91.72	0.100000
10	90.41	90.49	90.43	0.080000

	Operator 3			
Sample no.	Trial 1	Trial 2	Trial 3	Range
1	89.78	89.85	89.95	0.170000
2	90.68	90.63	90.74	0.110000
3	93.59	93.70	93.71	0.120000
4	91.66	91.84	91.91	0.250000
5	92.93	92.96	93.04	0.110000
6	89.04	89.04	89.20	0.160000
7	91.74	91.88	91.90	0.160000
8	93.11	93.13	93.19	0.080000
9	91.65	91.62	91.72	0.100000
10	90.41	90.49	90.43	0.080000

TABLE 12-5. Results of GR&R Study of Controlled Impedance Study

Part Id: 90–128	Upper tolerance: 99
Part number:	Lower tolerance: 81
Characteristic: Impedance	Tolerance spread: 18.00000
Gage Id: 90–128	Number of operators: 3
Gage name: TDR	Trials per operator: 3
Gage type: Impedance	Number of parts: 10

Average ($\bar{X}$)	Average ranges ($\bar{R}$)
A. 91.51733	A. 0.134000
B. 91.42433	B. 0.210000
C. 91.51000	C. 0.186000

Average ($\bar{X}$): 91.48389   Average ($\bar{R}$): 0.176667

Calculated upper control limit: 0.454740

	Measurement unit analysis	% Tolerance analysis
Equipment variation:	0.537402	2.985569
Appraiser variation:	0.230764	1.282020
Equipment and appraiser:	0.584853	3.219184

Study performed by: Chris C. & John A.

Case Study 2

Certain printed wiring boards require controlled impedance circuitry. Impedance is measured using *time domain reflectometry* (TDR). Measurement analysis here is key at this point of intervention to assure accurate reporting of this critical customer requirement.

Ten separate labeled coupons are measured for impedance three times each in a random order by three operators who would normally perform the analysis. The coupons are of external lines of 60 ohms. The results are seen in Table 12-4. The numbers are entered into a GR&R software package, which automatically performs all calculations. The results are shown in Table 12-5.

Results: The combined variation of equipment and appraisers absorbs only 3.2% of the tolerance spread, assuring full confidence in this measurement system.

References

1. Wheeler, D. J., and Lyday, R. W. 1984. *Evaluating the Measurement Process.* Knoxville, Tenn: Statistical Process Control, Inc.

2. Duncan, A. J. 1974. *Quality Control and Industrial Statistics*, 4th ed. Homewood, Ill.: Richard D. Irwin, Inc.
3. Barrentire, L. B. 1991. *Concepts for R&R Studies*. Milwaukee, Wis.: ASQC Quality Press.
4. Grant, E. L., and Leavenworth, R. S. 1988. *Statistical Quality Control*, 6th ed. New York: McGraw-Hill, Inc.
5. Stewhart, W. A. 1931. *Economic Control of Quality of Manufactured Product*. Milwaukee, Wis.: ASQC Quality Press.
6. ANSI/IPC-PC-90. 1990. *General Requirements for Implementation of Statistical Process Control*. Lincolnwood, Ill.: Institute for Interconnecting and Packaging Electronic Circuits.

13

The Use of Information System to Enhance Concurrent Engineering at MA/COM Omnispectra

Charles Ward
MA/COM Omnispectra

MA/COM Omnispectra is a $40 million manufacturing company of RF Connectors. Major customers include Texas Instruments, Hewlett Packard, Motorola, AT&T, Raytheon, and Sanders. The company was historically a primary supplier of connectors to the military. With the military contracts flat and declining, the company has focused on converting to become a commercial supplier. As part of the conversion process, they have changed their internal control systems to become a low-cost, high-quality, on-time supplier. The company is involved with concurrent engineering, total quality management (TQM), statistical process control (SPC), and working towards ISO 9000 qualification in order to compete in the European marketplace.

In order to achieve these goals, the information system was upgraded to speed up the development of new designs (Figure 13-1) and enhance the implementation of concurrent engineering. The three major goals of the information system are as follows:

1. Control the new product design and documentation by making all relevant information, such as specifications, material requirements, manufacturing procedures, costs, and special customer requirements, accessible to all departments. In addition, the information system will maintain the

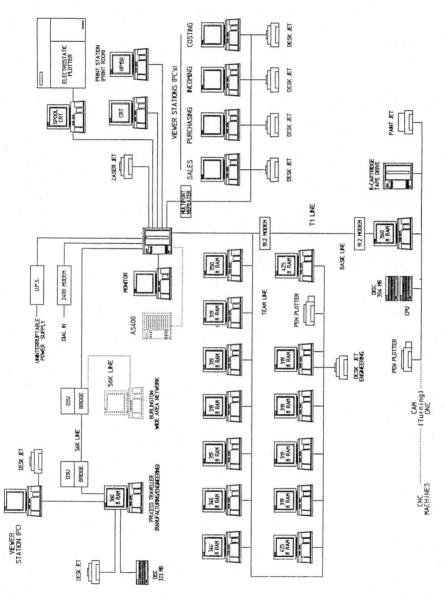

FIGURE 13-1. Omnispectra CAD/CAM system.

history and status of all engineering changes, to provide a legacy for making sound product decisions.
2. Facilitate the decision-making process of the concurrent engineering team by providing access to all members to product design and manufacturing information and by allowing for remote sign-off and document approval.
3. Increase the productivity of the concurrent engineering team members by providing on-line information access for procedural documentation of new products. The system itself is very efficient by having minimal programming and administration overhead.

HISTORICAL PERSPECTIVE

The Information System was launched originally in March 1988, with a 5-year plan to move into complete electronic connectivity. As a first step, the Omnispectra design/drafting room was converted into the CAD environment. The CAD system was planned to interface to the Materials Requirement Plan (MRP) system, with a history of 7 years in-house development in the company. At that time, the MRP system ran on the IBM System 38, which controlled all of Omnispectra's billings, materials, scheduling, planning, and cost. Later, the MRP system was updated to an IBM AS-5400. Design and engineering was performed on drafting boards, with an average of 10 days to several weeks to complete a part design.

A 2-D CAD, system consisting of four HP ME10 CAD systems, running an HPUNIX operating system was acquired initially. The system grew to fifteen stations and five viewer stations (see Figure 13-1). The viewer stations are low-cost, compatible personal computers, with 80386 microprocessors with 40-megabyte disc drives, and are connected to a low-cost graphical personal printer for obtaining hard copies.

In the MRP system, the costs are tabulated by the piece parts according to actual costs incurred, which are monitored continuously over time. These costs are established by current labor, material, and overhead rates. Any new part slated for quote is designed and documented before actual orders are received. These costs are extracted from the MRP system by searching for parts that are similar to the part being quoted. In the case of new designs, the manufacturing people assign an estimated cost for the parts.

To illustrate this methodology of assigning costs to new parts, Figure 13-2 is provided. The script function shows a drawing that is created in a user file, sent to a check file, and then sent to an approval file. All these files are signed off electronically in the MRP system. When the job is released, the drawings are sent electronically to a design file. This file is accessible by the manufac-

```
now processing the drawing log
/omnicad/logdir/drawinglog:4054-5005-50 moved from design for ECN Thu Nov 29 10:12:01
/omnicad/logdir/drawinglog:4054-5005-50 moved from check for changes Thu Nov 29 11:04:03
/omnicad/logdir/drawinglog:4054-5005-50 moved  to check for checking Fri Nov 30 12:13:21
/omnicad/logdir/drawinglog:4054-5005-50 moved from check for changes Tue Dec 04 11:05:34
/omnicad/logdir/drawinglog:4054-5005-50 waiting for approval Tue Dec 4 15:40:19
/omnicad/logdir/drawinglog:4054-5005-50 vaulted for ECN update Thu Dec 06 15:27: 04
/omnicad/logdir/drawinglog:4054-5005-50 vaulted ready for release Wed Feb 01 09:14:45
Find completed : Strike Return when through
```

FIGURE 13-2.   Script function for a drawing.

turing department concerned with material acquisition, tooling, and assembly of the new design. They extract the relevant information from the viewer stations, available at all location and connected through a local area communication network (LAN).

An interesting feature of this system is the lack of a system administrator. The system updates and relocates files automatically. For example, the procedure for engineering change notice (ECN) updates is handled by having the designer initiate an ECN script file. The drawing is forwarded to the initiator, but is locked out to any other requestor on the system until the ECN is cleared. This procedure will ensure that only information on the latest revisions is available.

There is wide access on the network for information requests that can facilitate different departments' jobs: Purchasing uses drawings obtained by on-line viewer stations to send purchase orders to their customers, thus eliminating manual requests to the print room and waiting time. Sales can examine quote requests, inquire about the status of current jobs, or respond to customer queries. Quality personnel have similar access to drawings for incoming inspections for parts that are manufactured either by outside suppliers or in house. Manufacturing uses the information from the viewer station to recall information necessary to document new parts or to solve day-to-day manufacturing issues. A list of option menus available to the different users is given in Figure 13-3.

DOCUMENTATION REQUIREMENTS AND CONTROL

Documentation needed for new parts include process travel documentation (Figure 13-4). This is a step-by-step instruction for the manufacture and assembly of products, in the proper sequence of operations. As the manufacturing engineers are creating this document, they can examine the parts and

```
MENU 1.

1. LIST APPROVAL
2. LIST CHECK
3. MORE BACKUP
4. LISTLOG
5. LIST OLD LOGS
6. LIST ECN'S
7. ps ref
8. who acq
9. SHOW COMMANDS
10.LIST COPYLIST
11.GO TO UNRELEASED LISTING MENU
12.LIST omnicad/viod
13.PRINT /BDF.LOG
14.GO TO DUPLICATE FILES MENU
15.QUIT MENU PROGRAM

echo, "please enter option number (1..15)"


MENU 2 UNREL

1. LIST unrel/APs
2. LIST unrel/contacts
3. LIST unrel/dash50/2xxx-
4. LIST unrel/dash50/others
5. LIST unrel/dielectrics
6. LIST unrel/housings
7. LIST unrel/layouts
8. LIST unrel/mfg drwgs
9. LIST unrel/misc
10.LIST unrel/not
11.LIST unrel/outline/2xxx-
12.LIST unrel/outline/other
13.List unrel/subassys
14.RETURN TO MENU 1
echo, "please enter option number (1..14)"


MENU 3 DUPLICATES

1. DUPLICATES IN MORE THAN ONE UNREL DIRECTORY
2. DUPLICATES IN MORE THAN ONE DESIGN DIRECTORY
3. DUPLICATES IN BOTH UNREL & DESIGN
4. DUPLICATES LAYOUTS IN BOTH UNREL & DESIGN
5. PRINT DUPLICATES IN MORE THAN ONE UNREL DIRECTORY
6. PRINT DUPLICATES IN MORE THAN ONE DESIGN DIRECTORY
7. PRINT DUPLICATES IN BOTH UNREL & DESIGN
8. PRINT DUPLICATES LAYOUTS IN BOTH UNREL & DESIGN
9. RUN DUPLICATES SCRIPT (runs automatically in cron at beginning of month)
10.RETURN TO MENU 1

echo, "please enter option number (1..10)"
```

FIGURE 13-3. Option menu available to information system users.

the tooling library on line. The process traveller is signed by the manufacturing engineer and then sent electronically to the Quality department, where it will be approved and returned to manufacturing. Final sign-off occurs electronically and then is folded onto the data base, where anyone accessing the system with a viewer station or a CAD station can electronically refer to this information on line.

SOFTWARE REQUIREMENTS ON PROCESS TRAVELER

OSI PART NUMBER -

CUSTOMER NAME -

CUSTOMER'S PART NUMBER -

PART MARKING

UNIT PACKAGING MATERIAL

TESTING

IN PROCESS INSPECTION

DESTRUCT SAMPLES REQUIRED

TEST DATA TO BE SUPPLIED

CERTIFICATE OF COMPLIANCE REQUIRED

SPECIAL PACKAGING

COMMENTS

DATE REQUESTED -

REQUESTED BY -

FIGURE 13-4. Process travel documentation.

An MRP board, which consists of engineering, quality, and manufacturing representatives, meets on a regular basis to discuss problems and how to resolve them. Information available from the viewer station, such as an ECN chronological history, manufacturing process, and current manufacturing status (such as the number of operations completed), will be instantly available to the group to facilitate resolution of problems.

This internal LAN system is connected to a wide area network system, so that some of the customers, suppliers, and other divisions of the corporation can log onto the system and get the drawings that they require electronically, without faxing or calling up the print room to see if that information is on file. They can also obtain the latest information and status of parts, quotes, and production schedules.

SYSTEM CONNECTIVITY

One of the problems of creating a hybrid information system such as this one is the lack of standards in the system components and operating systems. These were acquired and developed over time, with a lack of long-range planning for evolution of the system capability and connectivity. In most cases, the investment in current systems is too great in terms of training and procedural systems to allow for the purchase and conversion to new information systems. As an illustration, in this information system, there are three operating systems: HPUNIX for CAD, IBM for the AS400/MRP system, and MS-DOS for the viewer stations.

The connectivity between the system components is very complex, since information has to be transferred across several operating systems. Frequently, a search has to be initiated to locate a single drawing through as many as 10,000 drawings on file through several attributes. System response is less than 20 seconds.

The system also generates an important customer-oriented document called the outline drawing. It contains an envelope-sized drawing of the connecter, the components to build it, the materials to finish it, and the electrical, mechanical, and environmental specifications. This drawing is within easy access to the customer through networking by remote modem. Future enhancements of the system will enable the company sales force to satisfy customer needs immediately. They will be able to converse with the customer, obtain an on-line drawing of the latest revision of the product, and then electronically revise the drawing and store it back into the system. The design engineers could then give it a new number, recreate it, and then send a quote back to the customer.

The assembly procedures can also be improved for a multicultural work force. The system allows for graphically showing the assembly steps so that a worker who deficient in English can assemble connectors and cables without having to read the assembly instruction.

The material list is also provided by the system, with indentation to show the relationship of the different parts, their specifications, and any special features or operations required in production. The system is connected to the material lists generated by CAD and eliminates costly word processing and transcribing of material and part lists manually into the system.

The software requirement form is another document generated in the system. It relays the customer requirements to manufacturing engineers and quality. It details the history of the software charges and the testing, visual inspection, and mechanical requirements. The standard form for the software requirements is given in Figure 13-5.

ENGINEERING CHANGE NOTICES (ECNs)

Before implementing the information system, the engineering change notices (ECNs) were done by paper, distributed, and signed off through a very cumbersome system. With the networking inherent in the system, anyone sitting on a viewer station can initiate or query an ECN. Access can be done through part number or ECN number. Information available about the ECNs is shown in Figures 13-6 and 13-7.

With the two systems (MRP and CAD) connected together, the manufacturing engineer can develop new products in his or her process traveller environment, looking at the assembly drawing and the software requirements, and going into the MRP system to look at any ECNs, material prices, volumes, deliveries, or status on line. The traveller generation process averages 5 minutes, as opposed to the 30 minutes previously required in the manual system, with much greater quality and accuracy of the document.

RESULTS OF THE INFORMATION SYSTEM INVESTMENTS

The information system at Omnispectra fulfilled its promise of enhancing the new product development process, by reducing the time-to-market and increasing the quality of the design and manufacturing processes. Through easy access of relevant information to all members of the concurrent engineering team and providing for the framework of on-line documentation generation and approval, new product designs were completed in much shorter time, while significantly reducing revisions and engineering changes. The results are shown in Figures 13-8 and 13-9.

Since 1989, when the system was installed, new connector designs have increased by 50%. Jobs that used to take 10 days can be completed in 5 days, 90% of the time. At the same time, ECNs decreased from a high of 2,015 in 1988 to a low of 662 in 1992. The ECNs attributed to engineers making mistakes dropped from 35% in 1990 to 17% in 1992, and the number continues to drop sharply.

The investment in such an internally developed system is modest compared to the cost of commercial systems available today. In addition, the retraining of the employees is minimal, since the system builds on existing procedures and components. The total information system cost was $315,000 for hardware acquisition, and the payback for this investment was estimated at less than 2 years.

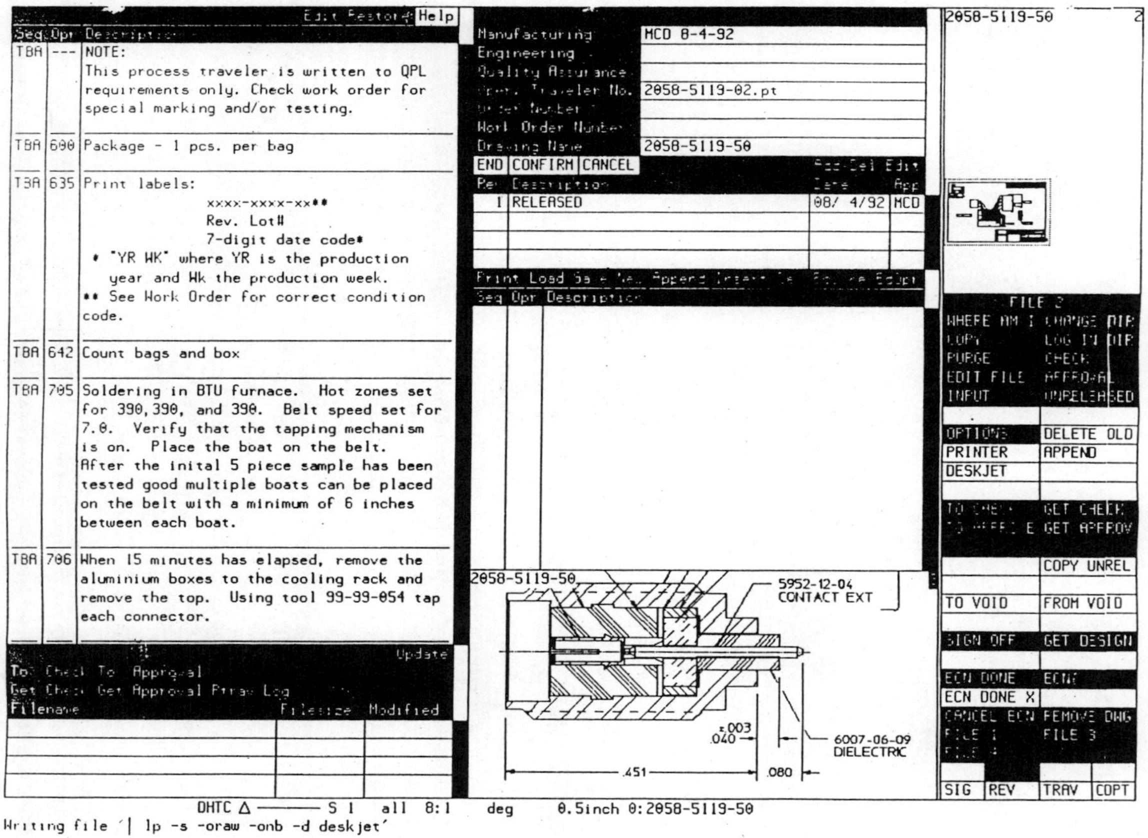

FIGURE 13-5. Software requirements form.

The Use of Information System   235

```
12/10/92              PRODUCT MANAGEMENT SYSTEM              WAL FPM 190
14:42:22          E. C. N. LOGGING HISTORY MAINTENANCE       DPM 066


    ECN #      CREATE      RELEASE      CLASS    REASON (A/C/D/E/M/Q/W)
    90-1162-2  12/10/92    0/00/00      2                E

                                                 REVISION
    PART #         DESCRIPTION     MAJOR         MINOR         P/L
    9999 5002 09   OSN    1          1             1           12
```

LETTER VALUE ECN

A - Add on (Drawing Maintenance - Draft)

C - Customer Change

D - Design Change - Correct old Designs to Dec 1988

E - Design Change - Correct Design Engineering Errors Jan 89 to present

M - Manufacturing

Q - QA Change

W - Watertown Manufacturing - Dimension and tolerance change

FIGURE 13-6.   ECN logging history form.

```
12/10/92              PRODUCT MANAGEMENT SYSTEM              WAL FPM 190
14:42:22          E. C. N. LOGGING HISTORY INQUIRY           DPM 066


    ECN #      CREATE      RELEASE      CLASS    REASON (A/C/D/E/M/Q/W)
    90-1162-2  12/10/92    0/00/00      2                E

                        T E X T  D E S C R I P T I O N
    9999 - 7885 -00  -02
    CHANGE FROM 2001-07-02 TO:    3001-07-03

    B/M'S -60'S, OUTLINES :
    MAJOR CHANGE TO 2012-11-02
    CHANGE OPERATION BLOCKS -50 PICTORIALLY
    DELETE : 3000-28-05    RETAINING RING
```

FIGURE 13-7.   ECN logging history inquiry.

236   Successful Implementation of Concurrent Engineering

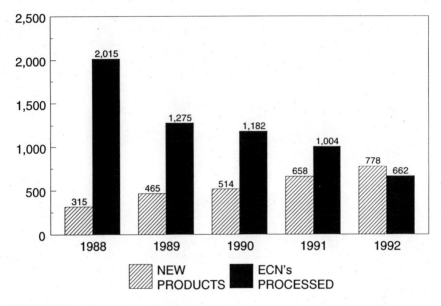

FIGURE 13-8.   Impact of the information system.

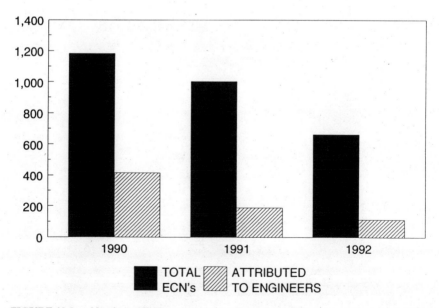

FIGURE 13-9.   Number of ECNs attributed to engineers.

The risk associated with an internally developed information system is the maintenance and upgrade of the system as the hardware and software technology moves forward. It is important to weigh these factors before embarking on a similar information system to support concurrent engineering.

14

Lessons Learned in Concurrent Product and Process Development

Susan Gross
Digital Equipment Corporation

This chapter describes lessons learned during the concurrent product/assembly process development effort for a computer peripheral product called A1. At the end of the effort, new technology had been introduced into the manufacturing plant that resulted in a robotic line capable of producing peripheral Product A1. Many aspects of the automation project were major successes.

However, at the end of the effort, the plant had several million dollars of equipment sitting idle, operators well trained in automation still building the current generation product by hand, and critical resources invested in equipment retrofits rather than production. In addition, the company was out of the A-family product business.

This chapter explores three observations about our company and its product/process development effort that affected the company's ability to benefit from the project investment. First, our process development *strategy* was product driven. Process decisions were based on product life cycles, without regards to the manufacturing plant life cycle.

Second, our process development *effort* was technology driven. A two-step and two-tier approach of technology first and operations second was used in developing and implementing the line. This over-the-wall approach changes the traditional picture of product/process investment risk over the life of a product.

Third, the plant used a departmental approach in supporting production. We learned that, for an automation line to provide the benefits it is designed to provide, a system view needed to be taken in defining the roles of engineering and operations resources supporting the A1 process.

Each of the three sections discusses our company practices relating to the development effort, explains the specific implications to the process decisions, and investigates what this means when applied across a company. Based on her experience as a process development engineer for A1, the author's goal is to communicate her understanding now that the outcome is clear and time has gone by, in hopes of broadening the readers' perspectives about product and process development projects involving automation.

CASE BACKGROUND

This case discusses the concurrent product and process development effort for computer peripheral product A1. This product was identical in function to the current generation product (A0), but was designed for automation in order to reduce its supply cost and to increase plant capacity.

The A1 printed circuit board was identical in function to the A0 printed circuit board. The main difference was that some components were changed so that automation could be used in assembling A1 circuit boards.

The manufacturing plant for A1 was the plant that was currently building A0. The current generation product had been in production for 5 years. The current process was intensively manual, and these were the skills that the plant had developed in its work force. The plant was using some of the oldest equipment in the company, which lent itself to a batch assembly process with a large investment in work in process and a large investment in a test-and-repair process. The company relied almost solely on this plant for this peripheral, so volumes were steadily high.

A process development team was formed to develop an automated process for product A1. Since this manufacturing approach was foreign to the plant, the engineering resources for this team were brought in from other parts of the company for this project. Part of the agreement for joining the team was that these engineers would only be responsible for developing the new, automated line. They would move on to other development projects once the line was being used for production. The development team was measured on two metrics: the per unit production cost for the projected volumes over the product life, and meeting the schedule for manufacturing readiness.

The robotic line for assembling A1 printed circuit boards is the focus for this case. A systems integration company was awarded the contract to deliver the line. This line was built at their site for debugging and testing and then installed in the manufacturing plant for acceptance testing and production. The process development team determined the requirements of the line and was heavily involved in the automation development.

A just-in-time philosophy was the basis for the design of the line. This encouraged queue sizes of one between operations and a serial pull system. It is very evident when there is a problem on a line with this design, because everything stops. For example, if one cell is stopped while its feeders are being loaded, the cell before it stops, since there is no place for the output to go, and the cell after it stops, since there is no input coming into the cell. The A1 process development team wanted to design in the practices of eliminating problems rather than having to cope with them. Since there was no place to buffer production, these practices were central to meeting throughput requirements. In support of this strategy, a distributed test strategy was used for the A1 process. Sensors were distributed throughout the process to detect component defects, assembly defects, and handling defects near their source. This was designed to enable the cause of problems to be determined quickly and thus facilitate their resolution and improvement in A1 process performance. Data from the *distributed test system* was part of the feedback mechanism vital to improving performance of the automation line.

The final layout of the line is shown in Figure 14-1. There were two identical parallel lines that placed components in the module. Each line was made up of one machine to automatically insert axial-leaded components, and three robots to insert parts such as DIPS and connectors. Components were tested in their feeder just prior to presentation to the robot. Defective parts were set aside in a tray. The robot stopped whenever two sequential parts failed. This alerted the operator that there was a problem that needed attention.

The equipment in each line was connected by conveyor, which limited the amount of work in process to one module between robots. After insertion, the conveyors from each line joined together. The modules were inspected using an automatic optical inspection system. The system was capable of detecting insertion defects, which were then repaired before soldering. The system was also capable of detecting subtle differences in the appearance of leads. Thus, process changes could be detected before they actually resulted in product defects.

Assembled circuit boards were then wave soldered, cleaned, tested, and moved to final assembly. No repairs were done after the board was soldered.

PRODUCT VERSUS PROCESS

The purpose of this section is to illustrate some of the characteristics of a product-driven organization, show the implications to the concurrent product/process development project, and discuss the ramifications to a company where similar practices are pervasive in the company culture.

In the context of concurrent product and process development, a product-driven strategy means that new processes are introduced strictly for new

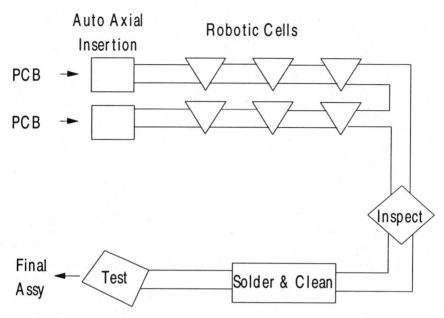

FIGURE 14.1. Layout of A1 circuit board assembly line. Two identical parallel lines are used to place components into the circuit board. Each line contains a machine for automatic insertion of axial-leaded parts and three robots for the other components, such as DIPS, connectors, and crystals. The equipment on each line is connected by conveyor, limiting the amount of work in process to one circuit board between each cell. After insertion, the conveyors join together for automatic optical inspection, wave soldering, clean, and test.

products. This approach means that a new product start-up foots the bill for the manufacturing project, which makes the new product start-up the primary customer for the process development team.

The practices that result from this strategy were well illustrated by the A1 project. In this case, not only was the new process developed and introduced for the new A1 product, this new process was the first substantive process investment in the plant. The A0 process had seen only minimal improvements in 5 years of operation. The antiquated equipment purchased to get A0 started was never replaced or even updated.

The company strategy was to fund projects based on immediate product payback. An internal product review noted that investing in a process to meet the needs of the plant rather than the needs of the product would require "investment in infrastructure," which was made "grudgingly."

All process decisions were considered in the context of the new product. Features that would have allowed the line to run future generation products were considered as not being affordable and did not pass the capital approval process. Included in these "extras" were a conveyor with adjustable width and an additional robot. This additional robot could be equipped with feeders for many different components. These features could have helped the plant transition from one product to another.

The development team designed a line the new product could afford for the projected volumes based on the capital appropriation approval. From then on, the focus was A1. Statements like "A1 is the only product that will ever run on this line" were accepted. The question of transitions was answered for transitions that happen during the life of the A1 product (pilot, revenue ship, volume, etc). There were several capacity studies done to show how the production needs of these different phases would be met under different assumptions about the readiness of the design of the A1 final system (the A1 system was to use a technology that was still in development) and the readiness of the automated manufacturing process. The risks were evaluated only within the context of the A1 product.

This product focus fostered an attitude of "Do it right the first time." The process development team was more concerned with avoiding the mistakes of the current A0 process than with reproducing the merits of it. The explicit approach was to make this process different from the A0 process and essentially start from scratch. Any merits of the A0 process that allowed it to produce 2,000 units a day were ignored.

The process development team's work was considered successful at the end of the project. The team had met their goal of introducing an automation line to the plant that was capable of producing A1 modules at the required rate and level of quality, as verified by the acceptance test. Unfortunately, the A1 product was cancelled abruptly because of factors other than the manufacturing process.

What about the plant? After completion of the process goal, the plant had not improved their situation. The new line was sitting idle, unable to produce the A0 product, when its production volumes were increasing to make up for the A1 cancellation. Plant personnel trained to support the automated line were not being utilized. The plant had to invest in a retrofit of the new line so that parts of it could be used to help meet A0 volumes. Even if A1 had not been cancelled, the plant would have had to invest heavily in a retrofit for A2 within a year.

The criteria for the company benefitting from the process development investment is balancing the needs of the product with the needs of the plant in terms of process flexibility. Three aspects are considered and illustrated for the *distributed test system* (DTS): cost, time, and performance.

DTS can be used to illustrate the relationship between system flexibility and system cost to the company. DTS was designed specifically for the A1 circuit board. It was adaptable for the A0 board only because the critical parts and the circuit schematic were the same on both modules. It was not adaptable to A2, the next generation product. A2 contained new functionality and thus would be a more complex design. Although the short-term cost was relatively low for a system of this nature, a DTS team would be needed every time a new part was introduced into the process, and new hardware and software would be required. There would be no way to take full and immediate advantage of the existing system for the next new product.

Figure 14-2 compares the cost of introducing a series of new products into a plant for two process development approaches. In the first case, a product-specific system, such as DTS, is developed for each new, related product. The introduction cost remains constant from one product to the next, as described

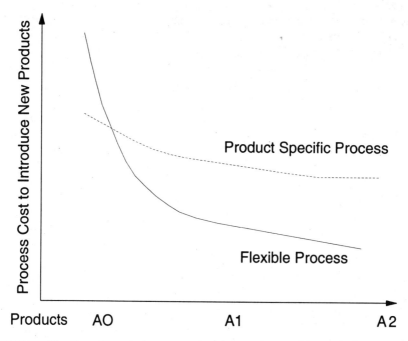

FIGURE 14.2. Cost of introducing new products into product-specific and flexible processes. This graph compares the cost to introduce new products into both product-specific and flexible processes. In the first case, a product-specific process is developed for each new, related product. The introduction cost remains constant from one product to the next. In the second case, a process designed to handle a rapid transition from one product to the next was developed for the first product. The cost for this flexible process is higher for the initial introduction. However, the cost of introducing a new product to this process decreases with each new, related product. A cost analysis that compares only the introduction cost for the first product may in fact lead to a more costly decision for the company over time.

above. In the second case, a system designed to handle a rapid transition from one product to the next was introduced for the first product. The cost for this flexible system is higher for the initial introduction. However, the cost of introducing a new product to this process decreases with each new, related product. At some time, the average cost over time of using the flexible system drops below the average cost for the product-specific systems. The cost analysis is very dependent on the actual systems, and no universal rule can be applied for using cost to justify one system over another. What we can learn from this comparison is that a cost analysis that compares only the introduction cost for the first product may in fact lead to a more costly decision for the company over time.

Yet, if another perspective is taken, one can draw a very powerful conclusion for the company. The Y-axis can be looked at as time-to-implementation, as in Figure 14-3. With time-to-market and product life becoming extremely short, there is no longer enough time to introduce a new system with each product introduction, regardless of system hardware and software costs.

This is precisely what happened.

The A2 product was needed sooner than originally expected. Because a major retrofit was required for A2 to be produced on the A1 line, the plant might not be able to deliver the process in time to meet A2 volume requirements. The company decided to have A2 built externally. And now the company is no longer in the A-family product business at all.

In addition to issues of cost and time, there is the issue of performance. A typical learning curve is shown in Figure 14-4. Benefits increase as we gain experience. The more we have to learn, the faster we do learn and the steeper the learning curve. The longer the system life, the more we can learn and the closer we get to maximizing the benefits of the system.

In actuality, learning curves for processes in a business are not continuous. As new products are introduced, there is always some break in performance. The extent to which overall performance is affected depends on whether the system is product specific or flexible.

Figures 14-5 and 14-6 show the learning curve for flexible and product-specific systems, respectively. For the flexible system, since the system has been planned for the products of the business, performance is only temporarily affected. In a short time, the previously attained level of performance is again reached. For the product-specific system, every time a new product is introduced, a new system is introduced. Every time a new system is introduced, the learning process starts all over again. For example, when a new DTS is used for the next generation product, some of the learning from the A1 system is lost, as it applies only to that specific product. Thus, we lose some of the benefits that we have invested in. As experience is gained on the new system, we move along the learning curve until the next product-specific

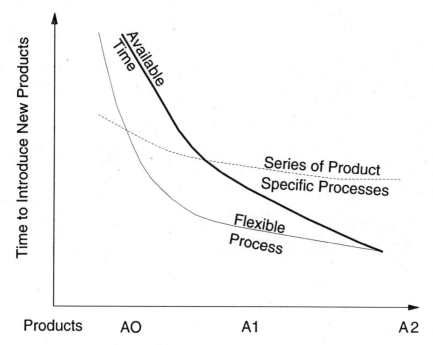

FIGURE 14.3. Time to introduce new products into product-specific and flexible processes. This graph compares the time to introduce new products into both product-specific and flexible processes. This is similar to the cost comparison, except that a rule can be applied universally in doing analysis based on time. As product life decreases, as shown by the available time curve, there will not be enough time to implement a new process for each new product. On this graph, the product-specific process is not ready in the time available for A2 development. The only way to have a process ready for a product is to have a flexible process that is designed to handle a rapid transition from one product to another.

system is put in place. Time pressures of the market make continuous improvement virtually impossible. With this series of ups and downs, on average the system operates at a suboptimal level over the life of the business (refer to Figure 14-6). Depending on the product life and system complexity, it is possible that a company may have a lower level of performance as it moves out in time through several product and process introductions.

To overcome this performance degradation with product-specific systems, steeper learning curves, as shown in Figure 14-7, would be required just to reach the previous performance level achieved, much less surpass it. In other words, employees and management must be more aggressive and more effective in terms of achieving process improvements, just to prevent a degradation in performance over several product introductions.

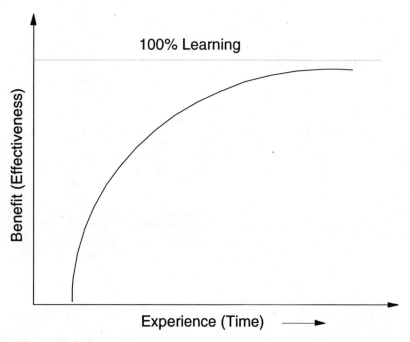

FIGURE 14.4. Learning curve associated with experience. Benefits increase as experience is gained. The longer the system life, the closer the company gets to maximizing its benefit (or performance) from the process.

Although company strategy and culture create the context in which process development decisions are made, product and process development programs can benefit a product-driven company. Fortunately, some of the needed improvement can be brought about by broadening the perspective of process development engineers and managers and arming them with lessons from process development projects such as this. There are two customers to every process: the product and the plant. Process investment decisions affect both customers and, ultimately, the viability of the business.

TECHNOLOGY VERSUS OPERATIONS

The purpose of this section is to describe a technology-driven company, show how these values affect the choices made in the automation project, and show how this increases the risk of product/process development projects to the company.

Not only did the company efforts center around the design of a new product, the company roots and organizational structure definitely favored

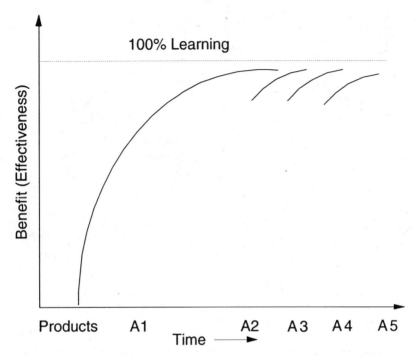

FIGURE 14.5. Learning curve for flexible process over several product introductions. Figure 14-5 shows the learning curve that a company is more likely to experience than the theoretical continuous curve. In reality, when each new product is introduced, there will be some setback in performance. With a flexible process, this will be quickly surpassed and the previous benefit level achieved. Performance is increasing over time.

developing technology over the application and utilization of that technology. As with the product focus, the more related to engineering that a function was, the more important it was and the more likely it would be to receive adequate funding and skilled resources. Designing a new process and introducing new technology and equipment was considered more valuable than improving an existing process by implementing more complete and thought-out operational strategies.

At the start of the A1 automation program, a process development team was formed to design and introduce the process to the plant. This team was made up of technologists brought in from within the company, but from outside the plant and outside the business environment. There are two important points here. First, the goal was to strengthen the plant by introducing a new process, not by improving the current process. Consider the jump in technology and skills required to run the automated A1 line over the

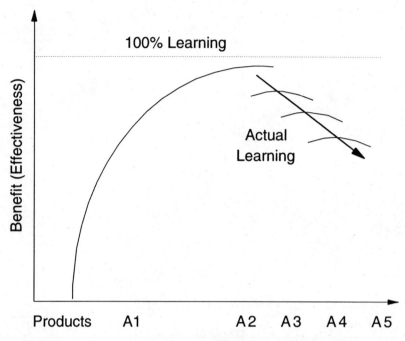

FIGURE 14.6. Learning curve for plant introducing product-specific processes for each new product. Every time a new product is introduced, a new process is introduced into the plant. Every time a new process is introduced, there is a significant setback in performance, as resources must focus less on production performance and more on new process developments. With this series of ups and downs, on average the plant performs at a suboptimal level over the life of several products. As the product life decreases, a potential reality is that a plant is not able to attain the previous level of performance before the next product and process introduction. In this case, the plant lowers its level of performance over time.

current manual process. Had a continual process improvement effort been supported, this jump would not have been so extreme, allowing resources from the plant with more operations experience to contribute to the project.

Second, by definition and as a contingency for joining the team, the development team was not to be responsible for the new line in volume production. The shared perception was that they were valued more by being technologists on development projects. In a similar manner, the plant support engineering groups were not responsible for the line until it was installed and accepted in the plant. These resources operated in a reactive, fire-fighting mode, as they responded to the day-to-day needs of A0 production. This did not allow them any time to focus on future processes. The levels of development and ongoing support resources across the life of the product are shown in Figure 14-8. The line was developed outside the business environment, to

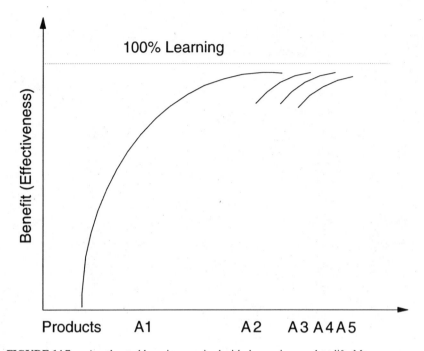

FIGURE 14.7. Accelerated learning required with decreasing product life. More aggressive process improvements (seen as steeper learning curves) in the plant are needed to prevent degrading performance if a product-specific process approach is used. Furthermore, in plants where the line is developed and installed by one team and handed off to another for production, this improvement must take place during the time when engineering ownership of the line is changing.

be integrated later. The priorities were technology introduction first and operational considerations second.

The activities of the process design team and the production operations team were based on this set of priorities. Concentric circles, as shown in Figure 14-9, can be used to describe the relationship between these activities. On the A1 project, process development concentrated on the inner part of the circle, the technological aspects of the process. They were concerned with the robot grippers, the algorithm for inserting components and clinching leads, and the order of picking and placing parts. The question for the process development team was, "For a good part on a good board, can the robot place the part and can the leads be clinched properly in the required time limit?" Process support was then forced to cope with this line in an actual production environment where all parts are not "good," where parts need to be brought to the line and loaded into the feeders. The operators worked to be as

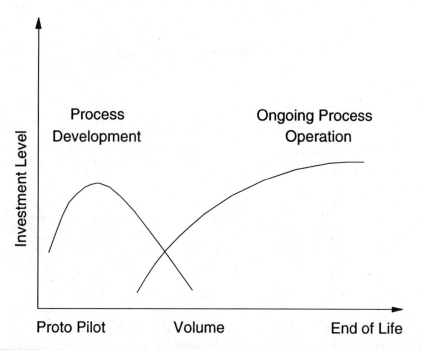

FIGURE 14.8. Levels of process development resources and ongoing operations resources at different stages of product life. Technology resources were responsible for designing and introducing the line into the plant. During the ramp-to-volume production, technology resources were phasing out of the A1 process, and ongoing operations resources were phasing in.

effective and efficient as possible in operating the line, but they had to do this within the constraints of its design. The operational activities in the outer layers of the circle were not inputs to the design of the line.

The result of this two-step approach for the A1 line was design errors that allowed only suboptimal process performance to be attained when the line was operated in an actual production environment.

For example, one of the first questions the operators asked when they were learning to operate the line was, "How do I reload the feeders?" On all the cells, the feeders were enclosed within the cell guarding. For safety reasons, in order for the robot arm to move, the guarding doors had to be closed. Thus, in order to have access to the feeders, the robot cell had to be stopped.

Initially, while there was one person dedicated to running each robotic cell, the operators could load feeders in parallel, reducing the down time of the line. After the initial introduction of the line, the plan for staffing the production line did not allow for each robot to have a dedicated operator. Loading feeders would result in a decrease in throughput. The cells would

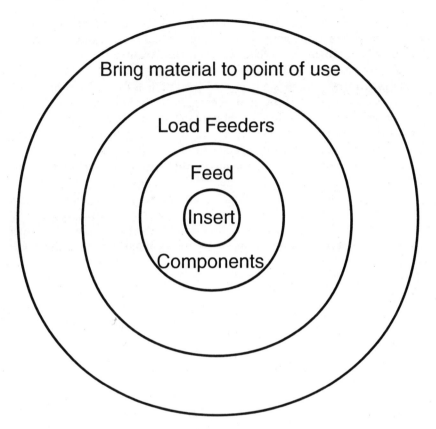

FIGURE 14.9. Process development and production operation activities. When technology is considered more critical than operations to the success of an automation line, the process design team tends to focus on the activities related to the bull's eye. A team can be enamored with the robot and its functions. the results of this are design choices that do not take into account the activities required to run the line in volume production. The other layers of the circle are not addressed until the line is run by the operators in their environment. By this time, there may be some design errors in the line that result in compromised performance.

have to be down serially, and the line would effectively be nonproductive during the entire loading procedure.

The operators and supervisors were creative in coming up with ideas for solving this problem. Although these ideas helped improve the downtime situation, the design of the line was a barrier to improving operations. For example, spare turrets were purchased, so a turret could be loaded off line and swapped onto the feeder quickly. This would have been an acceptable solution, except that the feeder had to be in one particular alignment for the

252   Successful Implementation of Concurrent Engineering

turret to be installed correctly. When this was discovered, the operators resorted to dumping the parts from one tube into an empty tube already on the turret. Although this helped alleviate the downtime problem, it also increased the number of defective components.

The principles basic to the design of the line had to be compromised if production goals were to be met. Because the issue of keeping feeders supplied with components was not thought through well by the technologists, the operators were at a disadvantage in performing their activities. The operational layers of the circle needed to be considered during the design. The approach of considering the layers of the circle is closely linked to the risk of process development investments to the company.

For a new product introduction in which there are significant product and process development costs, the company uses the curve shown in Figure 14-10 to illustrate the amount of risk associated with the different stages of product life. During pilot production and as the ramp to volume is started, the risk is

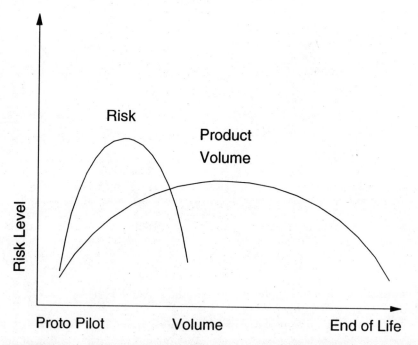

FIGURE 14.10.   Traditional assessment of risk levels over product life. This shows a curve that the company uses to describe the amount of risk associated with the different stages of product life for a typical product. Product volume increases slowly, from proto to pilot production. Risk increases as investment increases. The risk is highest when the plant is expected to produce at volume production levels.

Lessons Learned in Concurrent Product and Process Development 253

increasing rapidly. Significant investment in the product has already been made, and investment in the process is increasing quickly.

The risk is highest when the plant is expected to produce at volume production levels. This is due to the pressures on the plant and on the line. Not only will this likely be the first time that the equipment is run near capacity, but it may be the first time that material suppliers need this high a level of commitment. In addition, engineering changes ordered at this time have a tremendous impact on manufacturing. This two-tier, two-step development process significantly increases the risk of process investments for a company.

When the strategy is product driven and the development approach is technology driven, this picture changes dramatically (refer to Figure 14-11). The shape of the volume curve is sharper. Due to the dependency on the product, the start of pilot production may be delayed. Thus, the start of the curve is flatter. The date of first volume ship, however, is driven by factors external to the process and product development and may not be delayed. The second part of the curve must then be steeper to make up for this lost time, putting increased demands on the new line.

When operational issues are not addressed during the design of the line, they will surface when the line is operated in the plant during the ramp to volume production. This is an additional pressure that may hinder the actual performance compared to the planned performance, as we saw on the A1 line. This increases risk.

Second, ownership of the line is changing during the critical transition to volume production. Development resources are reducing their responsibilities and commitments, as planned, at the start of the project (refer to Figure 14-8). Operations resources are accustomed to a reactive, fire-fighting environment and will wait for a fire to break out before they take a role in running the process.

Not only did we have conflict that attracted the attention and drained the energy of everyone involved, but the progress that had been made on the line was halted. The communication channels so vital to the A1 production process were broken. The priorities for production support resources were not clear; thus, these resources were not dependable in resolving issues on the A1 line.

The combination of these conditions dramatically increases the risk of the line failing to meet the expectations and needs of the plant and of the product.

Understanding the consequences of our division between technology and operations and of our preference of technology over operations, it is apparent that part of the Japanese approach to manufacturing offers a solution to this problem. With the Japanese approach, developers of the line know up front

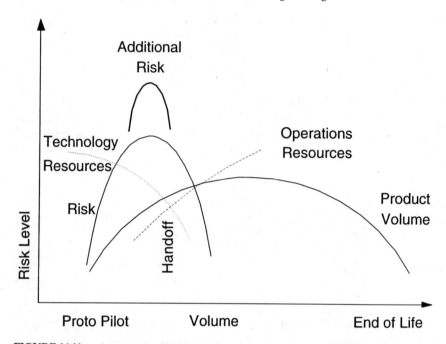

FIGURE 14.11. Assessment of risk levels for product and technology-driven processes. The curves in this figure better represent the risk associated with product- and technology-driven process development activities. The shape of the product volume curve is more jagged. Due to the dependency on the product design, the start of pilot production may be delayed. This results in a flatter curve, from proto to pilot, and a steeper curve, from pilot to volume, to make up for the lost time. When operational issues are not addressed during the design of the line, they will surface when the line is operated in the plant. This is an additional pressure that may hinder the actual performance compared to the planned performance. In addition, ownership of the line is changing hands at this same time. These factors compound the pressures brought on by the required rapid increase in volume. This results in potentially significant increases in the risk of the plant failing to meet production needs and, accordingly, the company failing to meet business needs.

that they will be responsible for operating the line in a volume production environment. They are considering these challenges throughout the decisions they make in the process design phase.

DEPARTMENTAL VERSUS SYSTEM VIEW

The purpose of this section is to describe an organization with a departmental approach, describe the challenges faced by the plant when automation was introduced into this context, and suggest how using a system view to redefine support roles will allow a plant to meet these challenges effectively.

Lessons Learned in Concurrent Product and Process Development

The plant, like many manufacturing plants, was organized by function. Quality engineering, vendor engineering, maintenance, process engineering, manufacturing engineering, and production were all responsible for solving their piece of the problem. Each "department" was measured on how well it carried out its function.

Although this allowed groups to focus their work and solve their problem efficiently, there was not a natural link between departments in the plant. In the manufacturing plant for A0, for example, departments did not report to the same staff-level manager. Moreover, the direct communication between employees of different departments was limited and not noticeably encouraged. The plant did not have one explicit overall process goal. A department worked to optimize itself, not the total process.

Production support was reactive. Support resources were called when a step in the process was unable to produce product. For example, the maintenance department was called when a machine went down. Process engineering was called when it was necessary to make a change to the equipment. Inspectors were responsible for detecting defects in the product. The goal of each of these departments was to "put the fire out" and be available to fight the next fire.

Although their involvement in the process was limited to when the process was broken, this departmental approach appeared to support the needs of production. When one step in the process stopped producing, the rest of the process appeared to be unaffected. There was enough work-in-process inventory that all the other functions could continue performing their functions even if a machine went down for a few hours.

The automated A1 line was vastly different, and three characteristics needed to be considered in supporting this line effectively.

First, the automated line relied heavily on the mechanical integrity of the parts. In the manual process, operators could make slight adjustments for bent or off-center leads and compensate for other mechanical problems by adjusting how they inserted the component. The feeder and robot could not make any of these adjustments; thus, the cell stopped whenever there were mechanical part problems.

Second, component packaging affected the ability of the line to run. The crystal was a good example of this. These parts came in a tube that was loaded horizontally between two supports in the feeder. Many of the tubes were too long to fit in the feeder. Obviously, the tube itself did not affect the functionality of the crystal or the circuit board. However, for this part to be inserted in a volume production environment, the tube shape and dimensions must be such that the tube fits in the feeder. Not only did the product functionality have to be considered in specifying parts, the process functionality had to be considered also.

Third, performance of the line depended on feedback from the product back to the process. As we have discussed, the production rate of the A1 line was more sensitive to cell downtime than was the A0 production process. In addition, the A1 line was more sensitive to defects. In the A1 line, any defective product found at test was to be scrapped. There was no time for major equipment adjustments during production or for producing product that could not be installed in an A1 system.

What happened when the automation line with these characteristics was introduced into the departmental environment? Initially, the line was operated like the A0 line. The operator tending each cell reverted to the practices he or she had always been measured on, maximizing throughput. Anticipating the effect of downtime of that and other cells due to parts problems and operational issues, he or she accumulated work-in-process buffers before and after his or her cell. This material had to be piled on chairs and on the floor near the cell, since the buffers on the line only held one board. The operator expected parts to jam and made tools to clear parts jams in the feeders without having to stop the line.

The resulting situation was a line connected only physically by conveyor. There was no feedback from the product to the process. Operators purposely prevented the line from stopping, from providing them with information about the problems that needed to be solved. In addition, there was no order to the work-in-process inventory, so, when a defect was found, an operator had no idea if this defect was a random occurrence or one in a series of the same defect. This resulted in the inability to relate what was happening on the product to what was happening in the process—the fundamental principle for improving the process so as to meet quality and throughput requirements over the life of the product.

Obviously, production was not managed this way for long. What resulted from this crisis was the realization that the A1 line had to be explicitly measured and operated as a system if it was going to meet the needs of the plant. The technology and mechanics of the line in and of itself could not guarantee the way that it was operated.

A different approach for determining the roles of support resources was needed. The departmental approach did not work. A system view had to be taken in defining the work that needed to be done. A system view requires that all activities and goals of manufacturing support resources be based on the needs of the process. There were three roles that changed dramatically on the A1 line once a system view was taken: the role of the maintenance technician, the role of the vendor engineer, and the role of the in-process inspector.

The role of the maintenance technician must be one of a process technician. He or she needs to be aware of what has been happening on the line in order to diagnose and fix a problem and then prevent it in the future. This process technician has a key role in moving the line along a learning curve. In several ways, it is this person who has the best opportunity to improve the productivity of the line.

The role of vendor engineering must change. The way that parts were judged for goodness was by a part specification, the set of characteristics required to make the part acceptable. This includes physical dimensions, warpage, and electrical/mechanical functionality. With automation, if parts are needed to function in a process as well as in the product, it seems logical that process metrics are needed for part goodness. The vendor engineer should be judged more on his or her ability to bring in parts that result in good output at the end of the line than to bring "acceptable" parts to the beginning of the line.

Their job also requires a more interactive relationship with suppliers. In the case of the circuit board, a partnership developed between the plant and the circuit board vendor. Although the circuit boards were within specification, a large fraction of boards were too warped for the automated equipment. Representatives from plant vendor engineering and manufacturing engineering worked with the board vendor so that they could understand the process better. The vendor was very receptive to our needs. In addition to sending an engineer to work with us, they ran experiments to determine the best material and process for flatter boards. The lesson here is that, just as the manufacturing engineer needs to be more aware of potential problems with parts in the process, the vendor engineer and the vendor need to be aware and much more involved in the process. The solution to process-material problems such as this required teamwork and integration across traditionally more detached organizations.

The role of the inspector must change. In the A1 line, an operator inspected the bottom side of the circuit boards with the use of an automatic optical inspection system. The inspection was done before the leads were soldered. There were two purposes for this inspection. First, since there was no repair after the solder and test, the inspection prevented assembly defects from being soldered to the board, which would have resulted in scrap after the test. Second, since A1 production was so sensitive to downtime, the purpose of the inspection was to detect changes in the process before they resulted in defects on the product. The inspection system worked by comparing each lead to the ideal of how the lead is supposed to look. The equipment discerns subtle differences; thus, drift of the assembly equipment can be detected early.

While the inspector can serve the first purpose without changing his or her role from that of an A0 inspector, that role changes significantly to serve the second purpose. The inspector is the person who can deliver feedback from the product to the process. He or she has the information to determine the source of defects and, thus, may be given the added responsibility of communicating that information and following through on the resolution.

The inspector must do more than inspect product. He must be able to analyze the data in real time, relate it back to the process, and take appropriate action to eliminate the cause. What used to be accomplished by a junior operator now requires an operator, technician, or supervisor with experience with all the process steps, equipment, and support processes.

The emphasis on fixing the process rather than fixing the product can be attained by employing a technician to be responsible for the inspection/feedback/resolution process. The point can be made very clearly by not allowing the technician to fix the defective product and allowing him or her to invest time strictly in eliminating the source of the defects from the inventory of potential problems.

In general, the metrics and reporting strategy did not encourage organizations or individuals to take a system view. What does this mean for a plant? Can I expand the roles of my employees without a massive investment in training and reorganization?

The answer to the second question is yes, and the primary reason is that the line indicates when and where there is a problem. The operators and other support resources can look at the line and know what is going on. This was impossible in the A0 process. They can then determine what actions they need to take. As long as employees understand that production depends on system performance and that all their efforts stem from what is happening in the manufacturing process, the roles take shape automatically.

This was demonstrated in the A1 process by the role that the operators and supervisor assumed in their jobs. They were the first to attempt to overcome the new challenges introduced with the automation line. They incorporated into their jobs much of what was typically done by off-line departments. They identified and grouped defective parts. They inspected tubes prior to attempting to load them to make sure they would work in the process. They brought manufacturing defects back to the source. They understood that they needed to work directly with the vendor or else have the vendor work directly in their process. They knew that they needed a maintenance technician to see the line when it worked and not just when it was already broken. By being involved in the process and looking at what it was showing them, the operators knew exactly what needed to be done to make the line successful. They knew this without any direction from engineering groups. They knew this because they were involved in and felt responsible for the overall process.

CONCLUSION

This chapter discussed how three characteristics of the company affected the way the product and process development project was carried out and explored some of the ramifications to a company where similar practices are pervasive in the company culture. In a product driven company, considering the life of the plant in addition to the life of the product is necessary for developing a plant–product–process relationship that benefits the company. Time-to-implement rather than cost-to-implement may be the factor that makes this consideration most meaningful.

The two-tier development process of technology over operations made the A1 line something that technical resources wanted to get rid of and something that operational resources thought could never work. The two-step process leads to design errors and suboptimal process performance during this transfer in the actual production environment. The combination leads to conflict and a dramatic increase in the investment risk to the company.

The A1 project showed us that the success of an automation line is as dependent on all aspects of the environment as it is on its design. Operating practices cannot be designed in; they must be managed in by changing reactive departmental metrics that focus on throughput to proactive process metrics that focus on reducing the inventory of potential problems.

It is hoped that this chapter has allowed readers to learn from the A1 development project. The goal was to provide a perspective that will enable projects similar to this to contribute to the viability of the business rather than to the demise of it.

Index

Index

A-1 matrix, 26
Acceptable parts. *See* Defects
Accuracy, 210–211
Activity-based costing (ABC), 45–48
Advance development, 1
Advanced Manufacturing Engineering Group (AME), 127
Affinity diagram, 22, 96
Algorithmic control, 82
Alternate Circuit Technology (ACT)
 history, 221
 product, 221
Analysis of operation, 191
ANOVA, 168, 179, 216
APOS. *See* Automated parts orientation system
APOS pallets, 154, 156
Appraiser variation (AV), 214. *See also* Reproducibility
Approved quality level (AQL), 208
AT&T, 226
Automated line, 238–240, 242, 248, 255
Automated parts orientation system (APOS), 151–152, 154
Automated tray changer (ATC), 152, 154
Automatic parts traying (APT), 152
Automation, 258

Banana jack, 50
Batch assembly, 239
Benchmarking, 75, 77, 207–208
Best in class, 120
Bill of materials (BOM), 129–132
Boothroyd-Dewhurst, Inc., 50, 109, 123

Breakthrough products, 1–3
Build it right the first time, 133
Bulk parts delivery, 151
Bull Information Systems
 Material Quality Engineering (MQE), 189–192, 196, 198, 201–202
 Supplier Quality Assurance Program (SQAP), 202
Business development plan, 8
Business Opportunity Statement, 99

CAD, 129, 132–133, 139–140, 150, 153, 228, 230, 232–233
CAE, 150
CAM, 105–108
Cause-and-effect diagram, 222
Chipcom Corporation
 business plan, 99–100
 concurrent engineering training, 109
 corporate management team (CMT), 89–90, 93–105
 culture, 88
 EDA, 106–107
 history, 89
 Manufacturing Council, 106
 NPDP, 90, 93, 95–96, 100, 101, 103–105, 109
 New Process Development Process. *See* NPDP
 PACT, 88–90, 98–102, 105, 109
  definition of, 89
  implementation of, 101–102b
  problems, 88
  product, 89
 Project Management Team, 90, 91, 93, 95–106, 108
 revenue, 89
 value statement, 90, 110
CMOS, 156

Index

Communication channels, 253
Communication with customer, 164. *See* Voice of the Customer
Communications, 80–83, 132
Communications with employees, 255
Competitive analysis team, 121
Competitive manufacturing lines, 163
Components, 139–140
Computer Integrated Manufacturing (CIM), 154
Computer simulation, 154
Concurrent engineering
   abilities, 95
   additional challenges, 61
   advantages, 44
   axioms for creating, 6
   benefits, 155
   boundaries, 88
   core team, 10
   cost penalties, 151
   communication, 125
   communications flow, 114
   correcting tooling deficiencies, 164
   cross-functional team, 10–11, 114
   definition, 1, 114
   early involvement, 123
   elimination of packing materials, 57
   enabling factors, 127
   functional units supporting, 8–10
   JITQC, 166–168, 174
   key processes, 13
   methodologies, 166
   need for, 72, 150
   NPI boundaries, 82. *See* New product introduction
   NPI plan, 72
   NPI Team, 72–74, 78–80
   NPI Team support, 74
   organizational model, 114
   part count, 59
   phase review, 13
   product driven organization, 240
   quality systems review, 13
   reasons for changing to, 99
   results, 62–63
   sample teams, 12
   solving technical problems, 9
   success of, 56, 113–114
   target dates, 123–125
   team concept, 10
   thorough planning, 56
   tools for
      control charts, 14
      design analysis, 15
      DFM analysis, 14
      engineering process, 14
      entity strategy, 14
      process capability, 14
      QFD, 15
      structured analysis, 14
      training, 153
Concurrent product and process development, 239–240
Concurrent project management, 11–13
Continual process improvement, 248
Coplanarity mean (COPMN), 168, 170, 171, 173, 179–180
Corporate goals, 163–164
Correlation matrix (A–3 matrix), 30
Cross-functional team, 46, 88. *See also* Team
Customer
   data collection, 23

feedback, 164, 166–167, 169, 170
interaction, 76
needs, 164, 232. *See also* Voice of the Customer
requirements, 22
customer service dedication, 199
Customer's voice, 107. *See also* Voice of the Customer
Cycle time, 164

Digital Equipment Corporation (DEC)
  corporate goals, 163
  corporate work flow data base, 165
  customer feedback, 176, 182–183
  distributed test system (DTS), 242–244
  investment in infrastructure, 241
  retrofit, 242
Defect avoidance, 165
Defect sensor, 240
Defective components, 252
Defects, 256–258
Deming, Dr. W. Edwards, 72, 206
Departmental approach, 255, 257
Design for manufacturability (DFM), 108–109, 122, 125, 144, 150, 153
  software, 109
Design for manufacturing and assembly (DFMA), 45–46, 48–50, 56. *See also* DFM
  redesign vs. new product development, 48–50
Design guidelines, 109, 140
Design managers, 107
Design/manufacturing relationship, 123

Design of experiments (DOE), 17, 207–208, 220
Design stability, 110
Development department, 9. *See also* R&D Engineering
Digital multimeter, 45
Distributed test system, 240
Do-it-right-the-first-time, 137, 242
Documentation, 143, 229
Duncan, 214
Dunnett's Multiple Comparison Test, 171, 180
Duplication of paperwork, 196

E-4 matrix, 32
ECN. *See* Engineering change notice or ECO
ECO. *See* Engineering change order or ECN
EDI capabilities, 106, 198
Electronic cross-reference part number, 198
E-mail, 95
Employee
  involvement, 167–169, 182, 258–259
  motivation, 164
  responsibilities, 191
End-of-life cycle, 173
Engineering change notice (ECN), 229, 231, 233. *See also* ECO
Engineering change order (ECO), 110, 120, 135, 252. *See also* ECN
Engineering task force, 169–170
Engineering vs. manufacturing, 128
EPA, 70
Equipment technology technicians, 174

Equipment variation (EV),
  212–213. *See also*
  Repeatability
Ethernet, 89, 94
Failure mode and effect analysis
  (FMEA), 35
Fault tree analysis, 35
FDDI networks, 89, 94
Firefighting, 248, 255
First customer ship (FCS), 117
Flagenbaum, Dr. Armond, 206
Flexible automation
  approach, 151
  equipment, 150
Flexible system, 244
Follow-on products, 1–2
Four P's, 192

Gage repeatability and
  reproducibility (GR&R),
  217, 219–220, 224
Gantt charts, 68, 78
Gate process, 138
Gates, 207
Global market, 150, 163
GR&R. *See* Gage repeatability
  and reproducibility
Greene Shaw Company, 189
  administrative rejects, 190
  competitiveness, 198, 201
  customer
    expectations, 198
    feedback, 199
    impact, 191
    needs, 202
  defects, 191
  employee morale, 201
  expanded computer system, 198
  long-range plans, 192, 196,
    198
  SOP, 199

  short-range plans, 192
  staff reduction, 199
  training, 23
  worksheets, 199

Hach Company
  development memo (DM), 74
  history, 70
  product, 7, 67
  Technical Training Centers
    (HTTC), 76
Hard automation, 159
Hewlett Packard Company, 2,
  226
Hewlett Packard Loveland
  market, 45
  product, 45
House of Quality (A-1 matrix),
  26, 28, 31–32

IBM, 93
Impedance, 224
Individual vs. team, 18, 66–68
Indoor connector, 19
  cable company advantages, 20
  customer benefits, 19–20
  specific needs of market of, 19
Information system, 226, 228
Inspection, 258
Inspection/feedback/resolution
  process, 258
Integrated product development,
  88
Internal process audits, 191
Intervention, 207
Ishikawa, Dr. Kaoru, 207
  Ishikawa diagram, 192
ISO standards, 144
  ISO 9000, 198–199, 203, 226
  ISO 9002, 135

Index   267

Japanese approach, 254
JITQC. *See* Just in time with total quality control
Juran, Dr. Joseph, 206
Just-in-time (JIT), 48, 135, 198, 240
Just-in-time with total quality control (JITQC), 164–166, 175–176, 181–183

K factors, 213–217
Kanban, 57, 174, 182
Kano model, 25
Kurtosis, 180

LAN. *See* Local area network
Lead tip average (LTAV1) 168, 170–171, 173–174, 179–180
Learning curve, 244–245
Linearity, 210, 220–221
Local area network (LAN), 81, 90, 229–230

MA/COM Omnispectra
  concurrent engineering, 226, 228
  history, 226
  product, 226
Makeup of team, 91
Malcolm Baldrige Award, 203
Management information system (MIS), 190–191, 198
Mann-Whitney test, 168, 174, 181
Manual assembly operation, 143
Manual process, 239, 255
Manufacturability assessment (MA), 14, 138, 140–141, 143
Manufacturing cells, 17
Manufacturing process development, 35, 36
Market requirements, 99

Materials Requirement Plan (MRP), 228, 231
Matrix of Matrices, 34
Matrix organization
  definition, 3, 5
  longer communication lines in, 4
  process focus in, 3
  process vs. product creation process, 4
Mean time between failure (MTBF), 12
Mean time to repair (MTTR), 12
Measurement system
  analysis, 208–210
  discrimination, 210–212
  effects of environment on, 220
  stability, 210, 220–221
Mentor, 107
Mercury Computer Systems, Inc.
  Advanced Product Planning Group (APP), 128
  operating philosophy, 129
  product, 126
Methodologies, 166
Metrology, 167
Milestone, 11–13, 132. *See also* Phase review
Milestone meetings, 12. *See also* Milestone
Modified products, 1
Motorola Inc., 226
Myers Briggs personality type analysis, 46

New product development process (NPDP), 120, 233
  strengths, 91–92
  weaknesses, 92
New product introduction (NPI), 72, 114, 128–129, 137, 139

New product module (NPM), 138–139
  process methodology, 129
  roles of team members, 117–119
  team, 114–115, 117
New products reevaluation, 94
New technology vs. improving existing process, 247
Non-value-added, 208
Northern Telecom
  corporate standards, 143–144
  corporate structure, 140
  design vs. manufacturing requirements, 149
  gate process, 138–139
  gate reviews, 139
  new product strategy, 138
  new product module (NPM) roles, 140
  tools, 144, 149
NPI. *See* New product introduction

Official product approval, 115
Off-site user training, 76
Original equipment manufacturer (OEM), 189
Outline drawing, 232

Packaging technology, 143
PCB, 16, 137, 141–142, 144
Pearson's Chi-Squared Test, 180
Penalty risk, 162
Pert charts, 68, 78
Petri Net, 66, 68, 75, 82
Phase review, 11–12
  project management, 139
Phase Two matrix in Four-phase system, 34
Plant investment, 241

Point-to-point wiring, 50
Polaroid Corp.
  interactive teams, 153
  SMART™
    calibration within, 156
    flexible autoline testing in, 156
    tooling design, 154
Process approach, 135
Process capability (Cp), 61, 150
Process development investments, 252
Process development team, 239–242, 247–250
Process development (technology driven), 238, 252
Process improvement, 17
Process investment, 246
Process support, 250
Process traveler, 229–230
Product and process development, 238, 246
  costs, 252
  product driven, 238, 252
  risks, 252
Product business plan, 98
Product creation process vs. matrix organization process, 4
Product creation using teams, 4
Product end-of-cycle, 114, 120
Product focus, 6
Product focus organization
  business units, 7–8
  marketing, 8
  product manager vs. project manager, 8
Product pipeline, 120–123
Product quality, 16–17, 166, 169
Product-specific system, 244
Product technical requirements
  from customer needs, 26
  prioritization of, 29

QFD matrix of, 28
  target values establishment, 28
Production department, 9–10
Production Management (PM), 128
Production operations team, 249
Productive process, 242
Programmable vision machine, 166
Project goals, 132
Project management, 5–6, 12–13
Prototype, 127
PWB, 126, 132–133. *See also* PCB, PWA

QFD. *See* Quality function deployment
  benefits, 37
  benefits of using, 47
  Competition analysis, 31
    comparison criteria, 32
    customer satisfaction, 31
  continuous improvement results, 196
  definition, 18
  design requirements, 32, 34
  impact of training, 36
  improved communications due to, 21
  key part features, 35
  matrix, 192
  new product development, 16
  output of, 47
  parts analysis, 34
  recommendations for application, 20
  standardization of use of, 40
  training
    direct participation, 21
    Four-phase approach, 21
    Matrix of Matrices, 21
  use of, 24
  weighting factor, 24

Quality, 206
Quality awareness, 97
Quality function deployment, 16, 45–47, 61, 75, 90, 100, 107, 190, 201. *See also* QFD
Quality improvement team, 202
Quality-minded culture, 207

R&R. *See* Repeatability and reproducibility
Raychem Corporation
  business units in, 17
  company culture in, 17
  products of, 16–17
  technologies of, 16
  Telecom Division, 18
  worth of, 17
Raytheon Company, 226
Repairability, 143
Repeatability, 210, 212–214
Repeatability and reproducibility (R&R), 210, 212, 215, 219–220
Reproducibility, 210, 214–220
Request for quotation, 135
Return on investment (ROI), 128, 161
Risk vs. reward, 3
Robotic assembly stations (RAS), 151–152, 154
Robust, 48
ROI. *See* Return on investment
RS/1 Graph and Statistics Program (BBN Corporation), 168

Self-directed work teams, 17
Serial new product development, 114
Serial pull system, 240

Serial vs. concurrent engineering, 72
Shewhart, Walter, 206
Simplified Timed Petri Net (STPN), 69, 83. *See also* Petri Net
Skew mean (SKMN), 168, 170–171, 173, 179–180
SLAM (computer simulation), 154
SMARTHUB, 93–94
SMART™ system (Sony Multi Assembly Robot Technology), 151, 154–156, 159
SMT packages, 165
Software application tools, 108
Software-based quality tracking, 198
Software-driven system, 152
Sony Corporation, 151–152, 159
SOP checklist, 196
SPC. *See* Statistical process control
Spread assembly machine, 155
Spread system, 155
Spreader assembly, 155
Standard operating procedure (SOP), 196
Statistical process control (SPC), 36, 135, 165–168, 206, 208, 210, 220, 222, 226. *See also* SQC
Statistical quality control (SQC), 36, 207. *See also* SPC
Strategic decisions vs. operational decisions, 66
Strategic product design, 88
Success factors, 127
Sun Microsystems
  history, 113
  product, 113

product approval form (PAF), 116
Product Strategy Committee, 116–117
Supplier
  communication, 136
  evaluation team, 134–135
  expectations, 136
  partnership, 134, 189
  relationship, 257
  selection, 136
Surface mount device (SMD), 126
Surface mount technology (SMT), 61

Target performance values, 28
Task list, 134
Task team, 97
Team
  approach, 139
  benefits, 66
  business team, 115–117
  charter statement, 21–22
  communication, 66, 78
  competitive analysis, 120–123
  cross-functional approach, 65–67
  cross-functional development
    formation of, 20
    members, 40
    training, 21
  decision making, 66
  design effort, 56
  dynamics of, 74
  effort, 66–67
  focus, 6
  formation of, 45
  functional team, 115, 117
  individual vs. team, 46, 66–68
  leadership of, 103
  make up of team, 115

management of, 103–104
members of, 45
players, 67
resources of team, 91
responsibilities, 91, 102
self-directed, 65–66
skills needed, 102
strategy of, 46
sub-teams within, 79
success, 77
suitability for product
  development, 67
team players, 67
team vs. committee, 66
training of, 46
turnover of, 74–75
work, 22
Teamwork, 139, 206–207
Technologists vs. plant support,
  248, 252
Technology driven, 246
Technology roadmap, 107
Technology vs. operations, 254
TESS (computer simulation), 154
Test and diagnostic process, 134
Test and repair, 239
Testability, 143
Test tools, 134
Texas Instruments, 226
Time domain reflectometer
  (TDR), 221, 224
Time to implementation, 244
Time to market, 233, 244
Token Ring, 89, 94
Tooling deficiencies, 164
Tools. *See also* under concurrent
  engineering
  flowcharts, 82
  process, 198
  used, 46
Total quality, 6

Total quality culture
  continuous improvements, 7
  focus, 6
  inhibitors, 7
  problem identification, 7
  team process, 7
Total quality management (TQM)
  17, 18, 190, 207, 201, 226
  training, 221
Total quality methods, 13
Toyota, 207
TQM. *See* total quality
  management
Track and via routing, 142
Traditional marketing functions,
  activities in, 5
Tree diagram, 22, 28

UL/CSA standards, 144
Upper control limit (UCLR), 219

Video conferencing, 81
Virtual enterprise strategy, 106
Voice of the Customer, 22, 47, 75

Walkman, 1511
Whats vs. hows, 24, 26
Work in process (WIP), 239,
  255–256
Work in progress, 164, 166, 168,
  174, 181–182. *See also* work
  in process
Worst case lead skew (WCL
  skew), 170, 177

Zenith Data Systems (ZDS), 190,
  202